AF183012

Wavelet Transforms and Their Recent Applications in Biology and Geoscience

Wavelet Transforms and Their Recent Applications in Biology and Geoscience

Editor

Jeff Winters

scitus academics

Wavelet Transforms and Their Recent Applications in Biology and Geoscience
Edited by **Jeff Winters**

ISBN: 978-1-68117-661-1
Library of Congress Control Number: 2016946528

© 2017 by
SCITUS Academics LLC,
616, Corporate Way, Suite 2, 4766,
Valley Cottage, NY 10989

www.scitusacademics.com

This book contains information obtained from highly regarded resources. Copyright for individual articles remains with the authors as indicated. All chapters are distributed under the terms of the Creative Commons Attribution License, which permits unrestricted use, distribution, and reproduction in any medium, provided the original author and source are credited.

Notice

Reasonable efforts have been made to publish reliable data and views articulated in the chapters are those of the individual contributors, and not necessarily those of the editors or publishers. Editors or publishers are not responsible for the accuracy of the information in the published chapters or consequences of their use. The publisher believes no responsibility for any damage or grievance to the persons or property arising out of the use of any materials, instructions, methods or thoughts in the book. The editors and the publisher have attempted to trace the copyright holders of all material reproduced in this publication and apologize to copyright holders if permission has not been obtained. If any copyright holder has not been acknowledged, please write to us so we may rectify.

Printed in United States of America on Acid Free Paper

Preface

Wavelet analysis is an exciting new method for solving difficult problems in mathematics, physics, and engineering, with modern applications as diverse as wave propagation, data compression, signal processing, image processing, pattern recognition, computer graphics, the detection of aircraft and submarines and other medical image technology. Wavelets allow complex information such as music, speech, images and patterns to be decomposed into elementary forms at different positions and scales and subsequently reconstructed with high precision. The main advantages of wavelet methods over traditional Fourier methods are the use of localized basis functions and the faster computation speed. Localized basis functions are ideal for analyzing real physical situations in which a signal contains discontinuities and sharp spikes. One of the main advantages of wavelets is that they offer a simultaneous localization in time and frequency domain. The second main advantage of wavelets is that, using fast wavelet transform, it is computationally very fast. Wavelets have the great advantage of being able to separate the fine details in a signal. Very small wavelets can be used to isolate very fine details in a signal, while very large wavelets can identify coarse details. A wavelet transform can be used to decompose a signal into component wavelets.

Wavelet Transforms and Their Recent Applications in Biology and Geoscience emphasize on the logical development of fundamental ideas and systematic treatment of wavelet analysis and its applications to a wide variety of problems as encountered in Biology and Geoscience. It is a valuable tool for practitioners and professionals.

Table of Contents

Chapter 1

Multiple Moving Objects Detection and Tracking Using Discrete Wavelet Transform

Chih-Hsien Hsia[1], Jen-Shiun Chiang[2] and Jing-Ming Guo[1]

[1] Department of Electrical Engineering National Taiwan University of Science and Technology, Taipei, Taiwan

[2] Department of Electrical Engineering Tamkang University, Taipei, Taiwan

1. INTRODUCTION

In recent years, video surveillance systems for the purpose of security have been developed rapidly. More and more researches try to develop intelligent video surveillance systems to replace the traditional passive video surveillance systems (Hu et al., 2004) and (Jacobs & Pless, 2008). The intelligent video surveillance system can detect moving objects in the initial stage and subsequently process the functions such as object classification, object tracking, and object behaviors description. Detecting moving object is a very important aspect of computer vision and has a very wide range of surveillance applications. The accurate location of the moving object does not only provide a focus of attention for post-processing but also can reduce the redundant computation for the incorrect motion of the moving object. The successful moving object detection in a real surrounding environment is a difficult task, since there are many kinds of problems such as illumination changes, fake motion (Cheng & Chen, 2006), night detection (Huang, 2008), and Gaussian noise in the background (Gonzalez & Woods, 2001) that may lead to detect incorrect motion of the moving object. There are three typical approaches for motion detection (Hu et al., 2004), (Jacobs

& Pless, 2008), and (Collins, 2000): background subtraction, temporal differencing, and optical flow. The background subtraction method detects moving regions between the current frame and the reference background frame. It provides the most complete motion mask data, but is susceptible to dynamic scene changes due to lighting and extraneous events. Therefore, it has to update the reference background frame frequently. The temporal differencing approach extracts the moving region by using consecutive frames of the image sequences. It is suitable for dynamic environment, but often extracts incomplete relevant motion object pixels. The optical flow method uses characteristics of flow vectors of moving objects over time to detect moving regions. However, most optical flow methods are with higher complex computation. Generally, the above three moving object detection methods are all sensitive to illumination changes, noises, and fake motion such as moving leaves of trees.

In order to solve the mentioned problems, several approaches for object detecting and tracking were proposed (Ahmed et al., 2005), (Alsaqre & Baozong, 2004), (Cheng & Chen, 2006), (Chen & Yang, 2007), (Collins, 2000), (Cvetkovic et al., 2006), (Hsieh & Hsu, 2007), (Hu et al., 2004), (Hu et al., 2009), (Huang et al., 2008), (Jacobs & Pless, 2008), (Liu et al., 2006), (Mckenna, 2000), (Sugandi, 2007), and (Tab, 2007). Video tracking systems have to deal with variously shaped and sized input objects, which often result in a massive computing cost of the input of images. Cheng *et al.* (Cheng & Chen, 2006) used discrete wavelet transform (DWT) to detect and track moving objects. The 2-D DWT can be used to decompose an image into four-subband images (LL, LH, HL, and HH). It only processes the part of LL-band image due to the consideration of low computing cost and noise reduction issues. Although this method provides low computing cost (low resolution) for post-processing and noise reduction based on the conventional DWT, the LL-band image produced by the original image size via two dimensions (row and column) calculation may cause high computing cost in the pre-processing. Especially they use the three-level low-low band image (LL_3) that does not only bring a great image size transfer computation, but also the slow motion of the real moving objects may disappear. After dealing with the background subtraction, Alsaqre *et al.* (Alsaqre & Baozong, 2004) used a local pre-process method to smooth the image with reducing noise and other small fluctuations. However, this approach is unable to reduce the post-processing computation. Sugandi *et al.* (Sugandi et al., 2007) proposed a method for detecting and tracking objects by using a low resolution image with the 2×2 average filter (2×2 AF), which is generated by replacing each pixel value of the original image with the average value of its neighbors and itself. They mentioned that the low resolution image is insensitive to illumination changes and can reduce the small movement like moving leaves of trees in the background. Although this method can deal with small movement, these low resolution images become more blurred than the LL-band image generated by using DWT.

To overcome the above-mentioned problems, we propose a method, direct LL-mask band scheme (DLLBS), for detecting and tracking moving objects by using SMDWT (Hsia et al., 2009). In DLLBS, we can select only the LL-mask band of SMDWT. Unlike the conventional DWT method to process row and column dimensions separately by low-pass filter and down-sampling, the LL-mask band of SMDWT can be used to directly calculate the LL-band image. Our proposed method can reduce the image transfer computing cost and remove fake motion that is not belonged to the real moving object. For objects occlusion, a new approach, characteristic point recognition (CPR), was proposed. Combined with DLLBS and CPR, it can have accurate object tracking for various types of occlusions. Furthermore, it can retain a better slow motion of objects than that of the low resolution method (Sugandi et al., 2007) and provide effective and complete moving object regions.

2. DISCRETE WAVELET TRANSFORM AND LOW RESOLUTION TECHNIQUE

Due to the imperfection of video acquisition systems and transmission channels, images are often corrupted by noise. Therefor, this degradation leads to a significant reduction of image quality, especially for the task that performs high-level computer vision, such as object tracking, recognition, etc. Before dealing with motion object detection, there are several methods for removing noises or fake motion and reducing computing cost proposed in the past several years. DWT (Cheng & Chen, 2006) and low resolution technique (Andra et al., 2000) are two important approaches, and are briefly described in the following sub-sections.

2.1. Discrete Wavelet Transform Method

Wavelet transform (Mallat, 1989) was proposed in the mid-1980s, and it has been used in various fields such as signal processing, image processing, computer vision, image compression, biochemistry medicine, etc. For image processing, it provides an extremely flexible multi-resolution image and can decompose an original image into different subband images including low- and high-frequencies. Therefore people can choose the specific resolution data or subband images upon their own demands (Hsia et al., 2009), (Mallat, 1989), (Ge et al., 2007), (Liu et al., 2006), (Ahmed et al., 2005), and (Tab et al., 2007).

A 2-D DWT of an image is illustrated in Fig. 1(a). When the original image is decomposed into four-subband images, it has to deal with row and column directions separately. First, the high-pass filter *G*and the low-pass filter *H* are exploited for each row data, and then are down-sampled by 2 to get high- and low-frequency components of the row. Next, the high- and the low-pass filters are applied again for each high- and low-frequency components of the column, and then are down-sampled by 2. By way of the above processing, the four-subband images are generated: HH, HL, LH, and LL. Each subband image has its own feature, such as the low-frequency information is preserved in the LL-band and the high-frequency information is almost preserved in the HH-, HL-, and LH-bands. The LL-subband image can be further decomposed in the same way for the second level subband image. By using 2-D DWT, an image can be decomposed into any level subband images, as shown in Fig. 1.

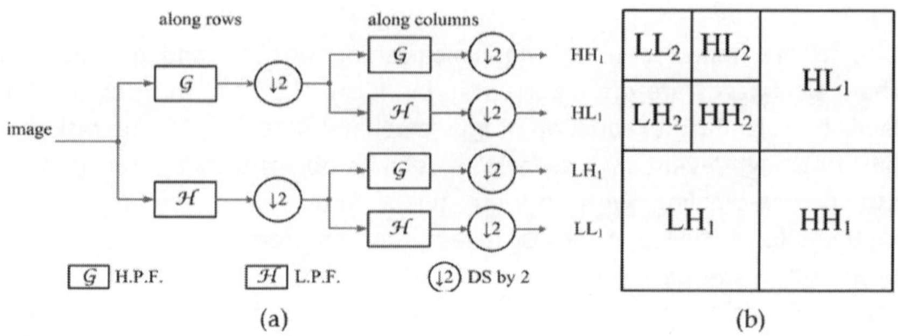

Figure 1. Diagrams of DWT image decomposition: (a) the 1-L 2-D analysis DWT image decomposition process, (b) the 2-L 2-D analysis DWT subband.

Cheng *et al.* (Cheng & Chen, 2006) applied the 2-D DWT for detecting and tracking moving objects and only the LL_3-band image is used for detecting the moving object motion. Because noises are preserved in high frequency, it can reduce the computing cost for post-processing by using the LL_3-band image. This method can be used for coping with noise or fake motion effectively, however the conventional DWT scheme has the disadvantages of complicated calculation when an original image is decomposed into the LL-band image. Moreover if it uses an LL_3-band image to deal with the fake motion, it may cause incomplete moving object detecting regions.

2.2. Low Resolution Method

Sugandi *et al.* (Sugandi, 2007) proposed a simple method by using the low resolution concept to deal with the fake motion such as moving leaves of trees. The low resolution image is generated by replacing each pixel value of an original image with the average value of its four neighbor pixels and itself as shown in Fig. 2. It also provides a flexible multi-resolution image like the DWT. Nevertheless, the low resolution images generated by using the 2×2 average filter method are more blurred than that by using the DWT method, as shown in Fig. 3. The average filtering is a low pass filter which denoises the image and performs restoration by the noise reduction spatial domain. It may reduce the preciseness of post-processing operation (such as occlusion and object identification), because the post-processing depends on the correct location of the moving object detecting and accuracy moving object data.

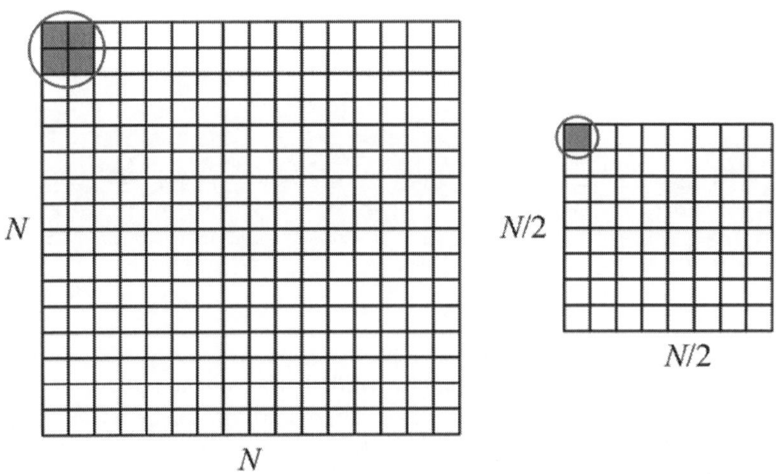

Figure 2. Diagram of the 2×2 average filter method.

3. DIRECT LL-MASK BAND SCHEME

In order to detect and track the moving object more accurately, we propose a new method called direct LL-mask band scheme (DLLBS) that is based on the 2-D integer symmetric mask-based discrete wavelet transform (SMDWT) (Hsia et al., 2009). It does not only retain the features of the flexibilities for multi-resolution, but also does not cause high computing cost when using it for finding different subband images. In addition, it preserves more image quality of the low resolution image than that of the low resolution method (Sugandi, 2007).

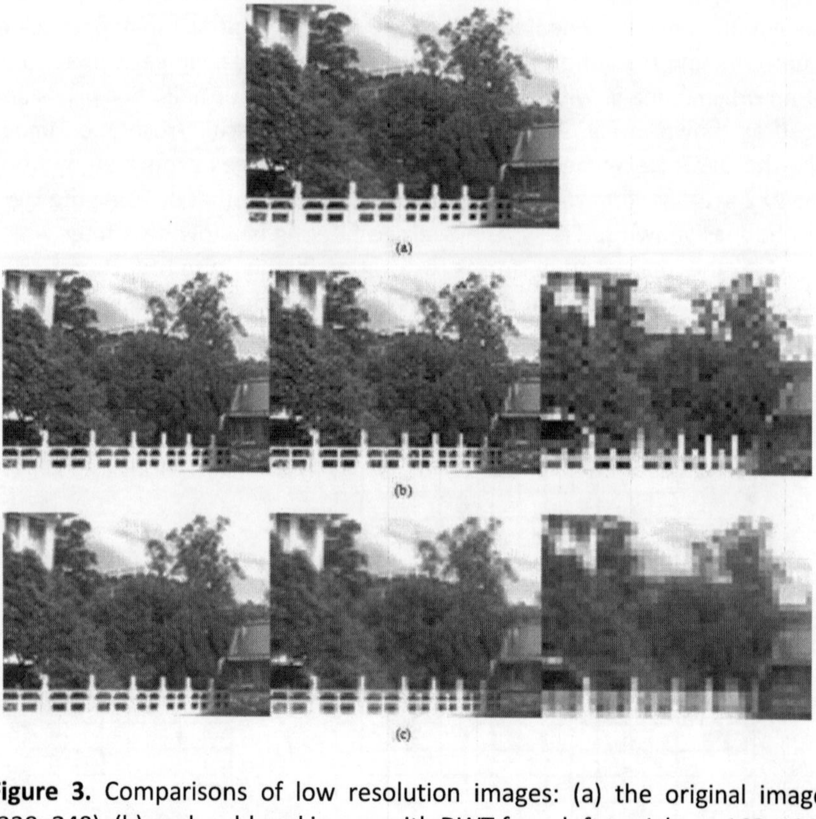

Figure 3. Comparisons of low resolution images: (a) the original image (320×240), (b) each subband image with DWT from left to right as 160×120, 80×60, and 40×30, respectively, (c) each resolution image with the 2×2 average filter method from left to right as 160×120, 80×60, and 40×30, respectively.

3.1. Symmetric Mask-Based Discrete Wavelet Transform (Smdwt)

In 2-D DWT, the computation needs a large transpose memory requirement and has a long critical path. The SMDWT has many advanced features such as short critical path, high speed operation, regular signal coding, and independent subband processing (Hsia et al., 2009). The derivation coefficient of the 2-D SMDWT is based on the 2-D 5/3 integer LDWT. For computation speed and simplicity considerations, four-masks, 3×3, 5×3, 3×5, and 5×5, are used to perform spatial filtering tasks. Moreover, the four-subband processing can be further optimized to speed up and reduce the temporal memory of the DWT coefficients. The four-matrix processors consist of four mask filters, and each filter is derived from one 2-D DWT of 5/3 integer lifting-based coefficients (Hsia et al., 2009). The coefficients of each subband mask are shown in Fig. 4, and the block diagram of the 2-D SMDWT is shown in Fig. 5.

1/4	-1/2	1/4
-1/2	1	-1/2
1/4	-1/2	1/4

(a)

1/16	-1/8	-3/8	-1/8	1/16
-1/8	1/4	3/4	1/4	-1/8
1/16	-1/8	-3/8	-1/8	1/16

(b)

1/16	-1/8	1/16
-1/8	1/4	-1/8
-3/8	3/4	-3/8
-1/8	1/4	-1/8
1/16	-1/8	1/16

(c)

1/64	-1/32	-3/32	-1/32	1/64
-1/32	1/16	3/16	1/16	-1/32
-3/32	3/16	9/16	3/16	-3/32
-1/32	1/16	3/16	1/16	-1/32
1/64	-1/32	-3/32	-1/32	1/64

(d)

Figure 4. The subband mask coefficients of (a) HH, (b) HL, (c) LH, and (d) LL.

3.2. Detection and Tracking Flow

The pre-processing flowchart of the proposed DLLBS moving object detection and tracking system is shown in Fig. 6. Frist, prior to color converting RGB data to YCbCr data (using Y data only). Basically we apply the double-change-detection method (Huang et al., 2004) to detect the moving objects. In order to decrease the holes left inside the moving entities, three continuous frames (F t_{-1}, F t, and F t_{+1}) are used in this system for detecting moving object mask. These three continuous frames are decomposed into LL_2-band frames (LL_2t_{-1}, LL_2t , and LL_2t_{+1}) by using SMDWT. After most of the noises and fake motions are moved into the high-frequency subband as shown in Fig. 7, it can proceed with the post-processing by employing these three LL_2-band frames. Binary masks, B t_{-1} and B t can be obtained by computing the binary values of these three successive LL_2-band frames (in between LL_2t_{-1}, LL_2t, and LL_2t_{+1}) and a threshold value T in (1).

$$B_{t-1}(i,j) = \begin{cases} 1, & \text{if } |LL_{2t-1}(i,j) - LL_{2t}(i,j)| > T \\ 0, & \text{otherwise} \end{cases},$$
$$B_t(i,j) = \begin{cases} 1, & \text{if } |LL_{2t}(i,j) - LL_{2t+1}(i,j)| > T \\ 0, & \text{otherwise} \end{cases}. \quad (1)$$

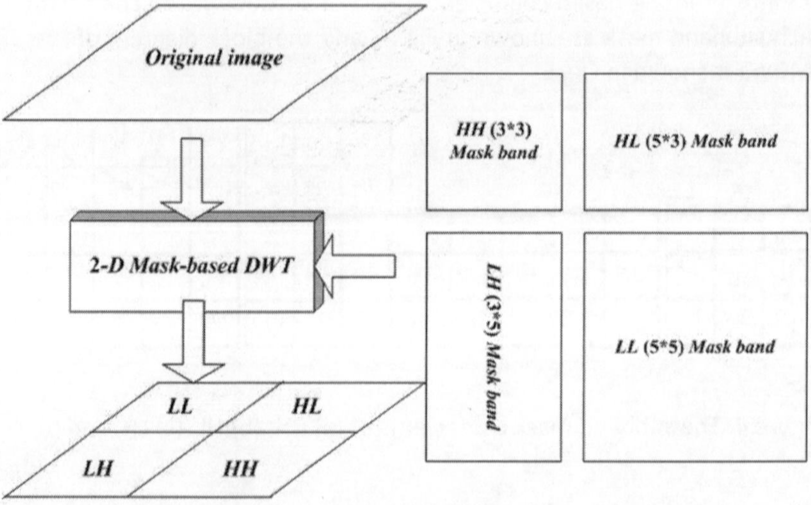

Figure 5. The system block diagram of 2-D SMDWT.

The motion mask (MM t) can be generated by using the union operation (logical OR) of B t_{-1} and B t. The function is represented as follows:

$$MM_t = B_t \cup B_{t-1}. \quad (2)$$

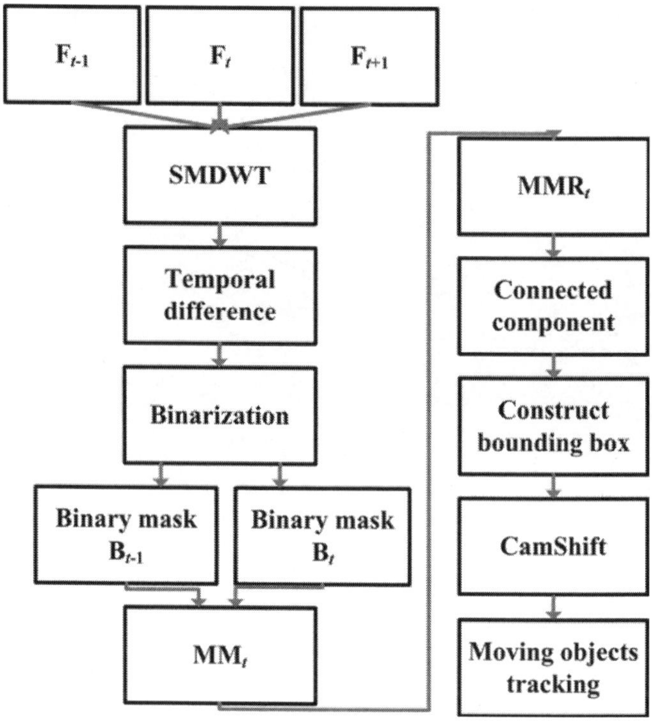

Figure 6. The pre-processing flowchart of the moving object detection and tracking based on DLLBS.

The holes may still exist in the motion masks, because some motion pixels are too tiny such that it causes error judgments as non-motion ones. In order to increase the motion mask (MM t) robustly, the morphological *closing* method (Hsieh & Hsu, 2007) is used to fill these holes. First, we apply the dilation operator for filling the middle of the isolated pixels that become related in the motion masks. It is defined as follows:

$$F_t(i,j) = \begin{cases} 1, & \text{if one or more pixels of the adjacent pixels of motion mask } MM_t(i,j) \text{ are } 1, \\ 0, & \text{otherwise.} \end{cases}$$

(3)

Then we apply the erosion operator for eliminating redundant pixels in the motion mask boundary as follows:

$$MMR_t(i,j) = \begin{cases} 0, & \text{if one or more pixels of the adjacent pixels of motion mask } F_t(i,j) \text{ are } 0, \\ 1, & \text{otherwise.} \end{cases} \quad (4)$$

Figure 7. After most of the noises and fake motions are removed using SMDWT (a) The original image, (b) LL_2-band image.

It scans eight neighbors of the motion mask MMR *t* image pixel by pixel from top left to bottom right (raster scan). After extracting the connected component, it obtains several moving objects. In this work, we utilize the region-based tracking algorithm (Cheng &. Chen, 2006), (Mckenna, 2000), and (Chen & Yang, 2007) to track the moving object motion.

Labeling is useful when the moving objects in the scene are more than one (The connected component labeling is then employed to label each moving object and track each moving object individually). The labeling of the components based on pixel connectivity (intensity) (Gonzalez & Woods, 2001) is obtained by scanning an image and groups, pixel by pixel from top left to bottom right, in order to identify the connected pixel regions by comparing the eight neighbors that have already been encountered in the scan. If the pixel has at least one neighbor with the same label, we label this pixel as the neighbor. The labeled moving objects are thus found, and then we extract the boundary of the moving object using rectangle box to track the moving object. For this reason, the bounding box is found according to its motion mask from the foregoing work. The bounding box is made by finding the minimum and maximum values of row and column coordinates of the motion mask. In order to track moving objects in the original image size, we have to transform the coordinate from the LL_2image size back to the original image size according to the spatial relationship of the DWT as follows:

$$O(i,j) = LL_n(i\times 2^n, j\times 2^n). \quad (5)$$

where $n = 0 \sim l$ and l is the number of level.

In the block-matching motion estimation, the motion vector is the displacement of a block with the minimum distortion from the reference block. The CamShift block-matching algorithm determines the motion vector by identifying a block with the minimum distortion from fast search strategies of the diamond-arc-hexagon search patterns in the search area (Chiang et al., 2008).

3.3. Occlusion Handling For Multiple Objects Tracking

In the post-processing, occlusion handling is a major problem in a video surveillance system. The most popular color space is the RGB color space (Hu et al., 2009). If the multiple objects bounding boxes are occluded, the object bounding boxes are merged into the occlusion bounding box. Here we propose a new approach for occlusion in multiple objects tracking, called characteristic point recognition (CPR). Fig. 8 shows the operation flowchart of CPR. CPR uses bounding boxes during pre-processing of DLLBS. For each tracked individual, the system will detect whether it makes occlusion with other object or group. It can obtain the RGB information from the video capture device directly to calculate the color information of the moving pixels. Owing to the information of moving pixels the size of the inter-frame difference image (1/16 of the original image) is with the central pixels.

To recognize every object, it uses the bounding box to find the characteristic point (CP). CP represents the central point of the bounding box as shown in the following equation:

$$Cs^q[n] = B_n\{(x_1, y_1), (x_2, y_2), \ldots, (x_q, y_q)\}, \quad (6)$$

where $Cs^q[n]$ is an array to store the CP of every object, *n* the label of the object, *q* the amount of CP, B *n* the bounding box of every object, and (x,y) the color information indexed by the position of CP. Therefore CP expresses the feature of the object. We would like to focus on each object bounding box in order to select one CP or more.

At first, the CP of every object is stored in the buffer when the first frame is input, and is regarded as the initial sample. In latter frames, the CP is matched with the sample. In other words, the CP of 1 to *n*matches with the CP of the sample as shown in the following equation:

$$Cd^q[n] = \mathrm{abs}\{((Cs^q[n]\text{- }Cm^q[N])_R,(Cs^q[n]\text{- }Cm^q[N])_G,(Cs^q[n]\text{- }Cm^q[N])_B)_N\}\ , \tag{7}$$

where $Cm^q[N]$ is a sample array to store the CP, *N* the label of the sample, and $Cd^q[n]$ the absolute values obtained from the difference between $Cs^q[n]$ and $Cm^q[N]$.

After the match step, $Cd^q[n]$ stores the sample *N* which is identical to the object *n* as shown in (8):

$$L[n] = N\ , \tag{8}$$

where L[*n*] is the label *N* to label object *n*. Therefore the object is recognized and labeled as sample *N*.

However, the objects of a frame may disappear or be occluded in latter frames. In order to hold the information of the object, the CP of the object has to be retained. Hence, we must know the object which has ever occurred when the object appears again in some frames. Because the CP may be changed by the environmental factors, the buffer has to be updated whenever a new frame is input in order to obtain the latest CP. If a new object appears, the CP of the new object should be added into the buffer to update the CP information. The CPR flowchart is shown in Fig. 8.

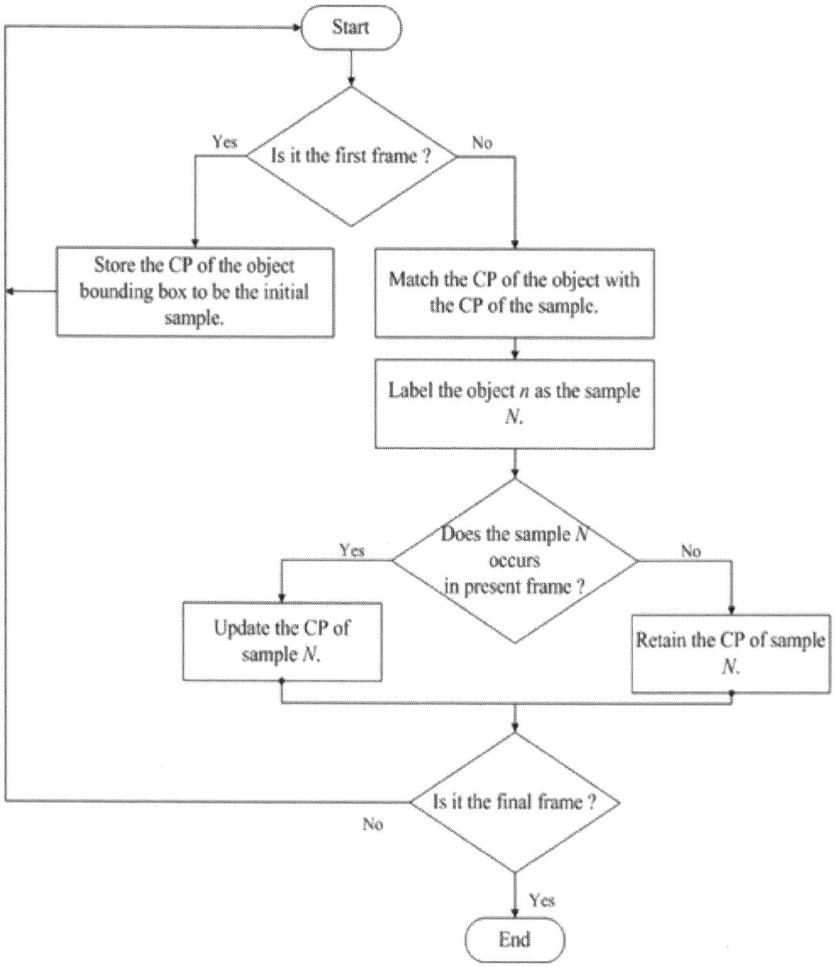

Figure 8.CPR flowchart.

4. EXPERIMENTAL RESULTS

In this work, the experimental results of several different environments including indoor (all day) and outdoor (all day) environments with statistic video system are demonstrated. The original image frame sizes are 320×240, 640×480, and the format of color image frame is 24-bit in a RGB system. We use all gray level frames from transferring the RGB system to YCbCr system for detecting moving object motion and utilize the LL_2 (for

320×240) and the LL_3 (for 640×480) image size of 80×60 generated by using SMDWT from the original image for our proposed moving object detection and tracking system. The experimental environment is set using Intel 2.83 GHz Core 2 Quad CPU, 2 GB RAM, Microsoft Windows XP SP3, and Borland C^{++} Builder (BCB) 6.0. BCB is chosen as the software development platform. The software includes verifying for algorithms and image process for the moving objects detection.

4.1. Dealing With Noise Issues

There are many kinds of difficulties such as illumination changes, fake motion, and Gaussian noise in the background. Different LL-band images including one-level, two-level, three-level, and multi-level LL-band images are used to deal with noises and compare their results. We suggest that a successful eliminating noise image has no other motion mask besides moving object motion masks, as shown inFigs. 9 and 10. Table 1 shows the average (Figs. 9 and 10) successful eliminating noise rate of each level LL-band image. The first row is in the indoor environment and the second row in the outdoor environment. Each level LL-band image has effective results when dealing with indoor noises like Gaussian noise produced by random noise and statistical noise. However, when dealing with the outdoor noise such as moving leaves of trees, the LL_1-band image has poor results because these outdoor noises sometimes are large that cannot be eliminated completely.

4.2. Moving Object Tracking

We consider it to have a complete moving object region if it is a successful work, as shown in Fig. 11(a). In Fig. 11(b), the moving object regions have only a part of moving object, and that will be treated as a failure tracking. Fig. 12(a) expresses the original frame without detecting and tracking moving objects. Without DLLBS technique many noise masks are tracked. However, even if the moving objects are tracked, those moving regions are fragmented, as shown in Fig. 12(b). By using DLLBS, the noises can be filtered out, as shown in Fig. 12(c). It still generates incomplete moving object regions by using LL_1-band image, because the relevance of these pixels in the LL_1-band image is deleted. When using a three-level resolution image to detect the moving objects, it generates incompletely moving object regions, owing to the LL_3-band image causing too many slow motions belonged to the moving object disappeared, as shown in Fig. 12(e). Finally, let us look at the results of the LL_2-band image in Fig. 12(d). Using the two-level band image has a better tracking region and also can cope with noises and fake motion effectively, as shown in Table 1.

Table 1. DLLBS: Direct LL-mask Band Scheme; [2]LS: Lifting Scheme; [3]2×2 AFS: 2×2 Average Filter Scheme; [4]DSS: Down-Sampled Scheme; [5]Accuracy rate: Success Tracking/ Original Sequency.The moving objects detection and tracking results.

Resolution	[1]DLLBS	[2]LS	[3]2×2 AFS	[4]DSS
Level	Accuracy rate	Accuracy rate	Accuracy rate	Accuracy rate
LL_1 (160×120)	99.54 %	99.54 %	99.07 %	98.15 %
LL_2 (80×60)	99.07 %	99.07 %	93.07 %	81.94 %
LL_3 (40×30)	95.83 %	95.83 %	86.11 %	63.89 %

[1]DLLBS: Direct LL-mask Band Scheme; [2]LS: Lifting Scheme; [3]2×2 AFS: 2×2 Average Filter Scheme; [4]DSS: Down-Sampled Scheme; [5]Accuracy rate: Success Tracking/ Original Sequency.

We use the 2×2 average filter scheme (AFS) in substituting the original DLLBS block system to demonstrate the moving object, however it is more blurred than the DLLBS technique. The accuracy rate of the successful object tracking with the 2×2 AFS are shown in Tables 1, 4, and 5. It is easy to perceive the contrasts between Tables 1 and 4 of any resolution image; the LL-band image generated by the DLLBS has a better successful ratio than the low resolution image generated by the 2×2 AFS.

Several experiments have been made to prove the feasibility of the proposed approach for moving object detection, tracking, and occlusion. We used an entry-level video camera and capture card to capture the test sequences in our campus (Tamkang University), and simulated several cases of condition for moving objects, such as signal object in day time (indoor/outdoor), signal object at night (outdoor), and multiple objects in day time (outdoor) environments. All the test sequences are stored as the Microsoft AVI format with raw file of resolution 320×240, 640×480, and frame rate of 30 fps as shown in Fig. 13.

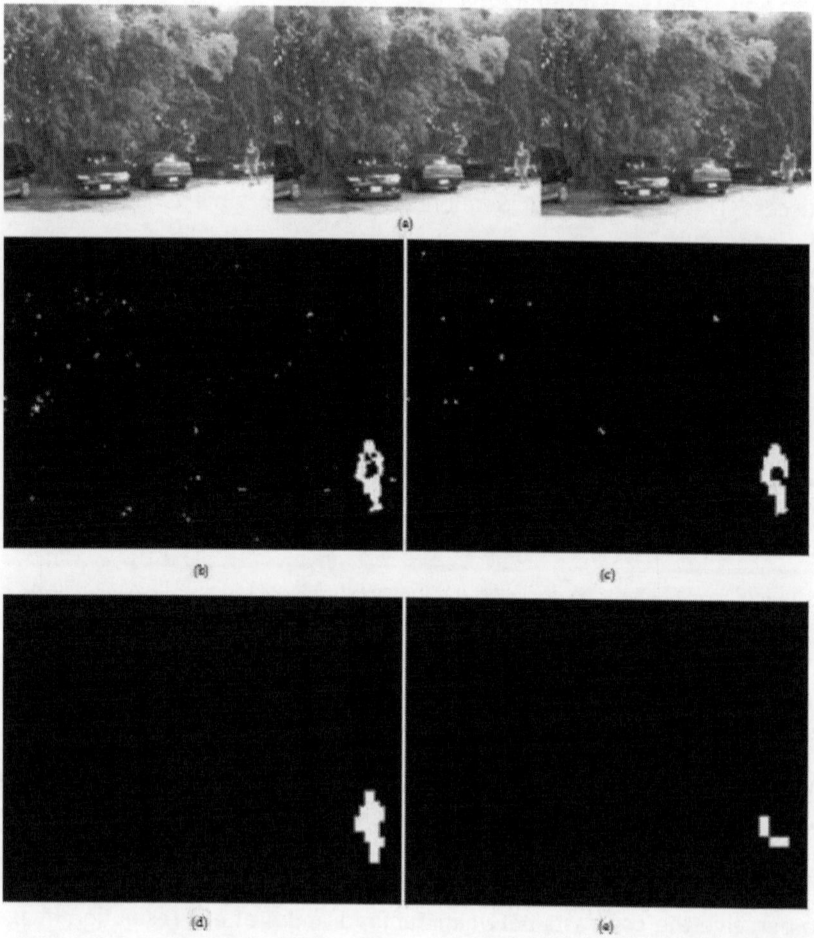

Figure 9. Moving object detection in the outdoor environment with fake motion: (a) the original image of three consecutive frames, (b) the temporal differencing results of the original image, (c) the temporal differencing results of the LL_1-band image, (d) the temporal differencing results of the LL_2-band image, (e) the temporal differencing results of the LL_3-band image.

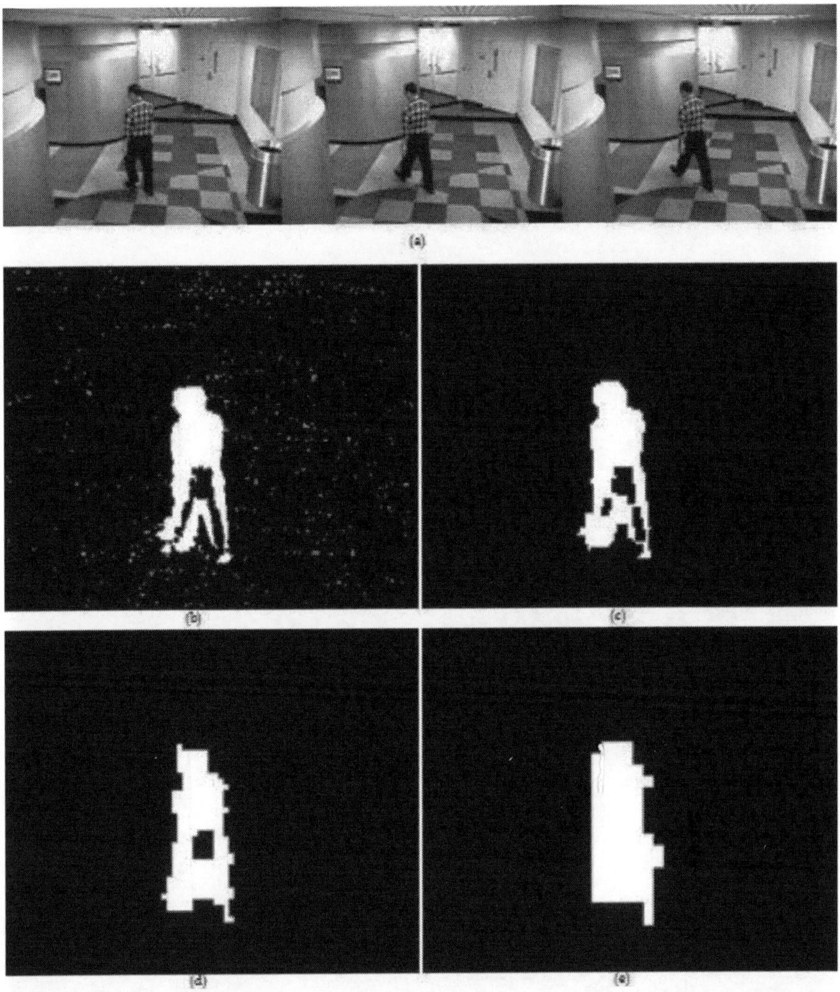

Figure 10. Moving object detection in the indoor environment with Gaussian noise (Gonzalez & Woods, 2001): (a) the original image of three consecutive frames, (b) the temporal differencing results of the original image, (c) the temporal differencing results of the LL_1-band image, (d) the temporal differencing results of the LL_2-band image, (e) the temporal differencing results of the LL_3-band image.

The choice of the threshold *T* is important. A too large value may lead to real targets missing; on the other hand a too small value may lead to noise binary images with pseudo features. The best threshold value also varies with the instantaneous illumination level. The threshold values of the best performance in different environments and DLLBS are listed in Table 2. According to the experiments, under day and midday time in outdoor for LL_2, the threshold values of 10, 14, and 16 were applied, some noisy pixels appeared. However, when the threshold value of 15 was applied, it outperformed all other threshold values.

Figure 11. Examples of (a) successful moving object tracking and (b) failure moving object tracking.

We established 16 test sequences at Tamkang University in different environments, such as day time, night time, rainy day, fast movement, slow movement, and occlusion, as shown in Fig. 13. Compared with other approaches (2×2 AFS and DSS), the DLLBS can obtain a good sparsity for spatially localized details, such as edges and singularities, as shown in Table

3. Because such details are typically abundant in natural images and convey a significant part of the information embedded therein, DWT has found a significant application for image denoising. From Tables 4, 5, and 6, we notice that some objects are not correctly identified in the test frame of the sequences. The wrong identification occurs in two reasons:

Figure 12. Results of tracking moving objects in various environments: (a) original frames without region-based object tracking, (b) original frames with region-based object tracking, (c) LL_1-band frames with region-based object tracking, (d) LL_2-band frames with region-based object tracking, (e) LL_3-band frames with region-based object tracking.

Table 2. The best threshold values, *T*, in different environments and DLLBS.

Resolution	Night in the outdoor	Day and Night in the indoor	Day and Midday in the outdoor
LL_1(160×120)	$T = 14$	$T = 20$	$T = 25$
LL_2 (80×60)	$T = 4$	$T = 10$	$T = 15$
LL_3 (40×30)	$T = 1$	$T = 6$	$T = 11$

Table 3. Features of various methods.

Environments	Fake motions	Low contrast	Reflection
DLLBS	Excellent	Excellent	Excellent
2×2 AFS	Excellent	Poor	Good
DSS	Poor	Excellent	Poor

The moving object just enters or leaves the scene.

Because the moving object is detected and tracked at the border of the scene, the extracted features of the moving object in the case cannot represent the moving object very well.

The moving object is slowing down.

In this issue, the temporal difference image of the object becomes smaller and loses its situation.

5. CONCLUSIONS

The direct LL-mask band scheme (DLLBS) for moving object detection and tracking is proposed in this work. It is able to detect and track moving objects in indoor and outdoor environments with statistic video systems. The proposed DLLBS does not only overcome the drawbacks of high complex computation and slow speed for the conventional DWT, but also preserves the wavelet features of the flexible multi-resolution image and the capability for dealing with noises and fake motion such as moving leaves of trees. In the real-word application, the experimental results demonstrate that the 2-D LL_2-band (for 320×240) and the 2-D LL_3-band (for 640×480) can effectively track moving objects by region-based tracking under any environments (day and night), as well as it can cope with noise issues. For occlusion considerations, we propose a new approach, characteristic point recognition (CPR). Combined with DLLBS and CPR, it can accurately track various types of occlusions. The DLLBS system can be extended to the real-time video surveillance system applications, such as object classification and descriptive behaviors of objects.

Table 4. Single moving object processing (without occlusion).

Pattern	DLLBS		2×2 AFS		DSS	
	Accuracy rate	Detection + Tracking	Accuracy rate	Detection + Tracking	Accuracy rate	Detection + Tracking
Sequence1	98.61 %	53.8 FPS	98.61 %	58.5 FPS	71.76 %	52.1 FPS
Sequence2	95.90 %	56.5 FPS	96.31 %	57.1 FPS	56.15 %	49.0 FPS
Sequence3	97.40 %	54.1 FPS	92.36 %	60.7 FPS	96.59 %	63.1 FPS
Sequence4	93.55 %	54.3 FPS	82.61 %	62.0 FPS	92.65 %	63.2 FPS
Sequence5	82.97 %	56.7 FPS	80.44 %	60.2 FPS	77.29 %	65.6 FPS
Sequence6	82.57 %	53.9 FPS	78.90 %	61.5 FPS	46.79 %	63.9 FPS
Sequence7	90.28 %	54.7 FPS	37.50 %	61.4 FPS	40.28 %	63.5 FPS
Sequence8	83.33 %	55.1 FPS	73.46 %	62.4 FPS	78.40 %	61.2 FPS
Sequence9	90.16 %	53.5 FPS	75.13 %	59.0 FPS	83.94 %	58.5 FPS
Average	90.53 %	54.7 FPS	79.48 %	60.3 FPS	71.53 %	60.1 FPS

Table 5. Multiple moving objects processing (with occlusion).

Pattern	DLLBS		2×2 AFS		DSS	
	Accuracy rate	Detection + Tracking + occlusion	Accuracy rate	Detection + Tracking + occlusion	Accuracy rate	Detection + Tracking + occlusion
Sequence10	92.94 %	56.7 FPS	88.61 %	60.6 FPS	82.92 %	57.6 FPS
Sequence11	89.98 %	54.8 FPS	83.67 %	60.9 FPS	92.60 %	58.7 FPS
Sequence12	90.43 %	53.9 FPS	79.79 %	60.8 FPS	82.98 %	55.9 FPS
Sequence13	90.00 %	52.6 FPS	86.67 %	59.7 FPS	75.56 %	63.7 FPS
Average	90.84 %	54.5 FPS	84.69 %	60.5 FPS	83.52 %	58.9 FPS

Table 6. Multiple moving objects processing (with occlusion).

Pattern	DLLBS		2×2 AFS		DSS	
	Accuracy rate	Detection + Tracking + occlusion	Accuracy rate	Detection + Tracking + occlusion	Accuracy rate	Detection + Tracking + occlusion
Sequence14	85.60 %	14.1 FPS	73.60 %	16.5 FPS	31.20 %	15.5 FPS
Sequence15	88.36 %	14.1 FPS	79.45 %	16.6 FPS	63.01 %	15.3 FPS
Sequence16	81.33 %	14.2 FPS	70.67 %	16.6 FPS	33.33 %	15.2 FPS
Average	85.10 %	14.1 FPS	74.57 %	16.6 FPS	42.51 %	15.3 FPS

Figure 13. Test sequences at Tamkang University: (a)-(k) are sequences 1-13 (320×240) and 14-16 (640×480); (a)-(c) show the single moving object in the outdoor; (d) single moving object in the indoor; (e) single moving object in the outdoor (fast movement to slow movement); (f) single moving object in the outdoor (fast movement); (g) single moving object in the outdoor (slow movement); (h) single moving object in the indoor (zoom-out to zoom-in); (i) single moving object in the outdoor (rainy day); (j) multiple moving object in the outdoor (occlusion); (k) multiple moving object in the outdoor; (l) multiple moving object in the outdoor (occlusion); (m) multiple moving object in the outdoor (occlusion); (n) multiple moving object in the outdoor (occlusion); (o) multiple moving object in the outdoor (occlusion); (p) multiple moving object in the outdoor (occlusion).

REFERENCES

1. J. Ahmed, M. N. Jafri, J. Ahmad, 2005Target tracking in an image sequence using wavelet features and neural network, IEEE TENCON, (November 2005) 1 6

2. K. Andra, C. Chakrabarti, T. Acharya, 2000A VLSI architecture for lifting-based wavelet transform, IEEE Workshop on Signal Processing Systems, (October 2000) 70 79

3. F. E. Alsaqre, Y. Baozong, 2004Multiple moving objects tracking for video surveillance system, IEEE International Conference on Signal Processing, 2August 2004) 1301 1305

4. D. Chen, T. , J. Yang, 2007Robust object tracking via online dynamic spatial bias appearance models, IEEE Transactions on Pattern Analysis and Machine Intelligence, 29 12December 2007) 2157 2169

5. F. Cheng, H. , Y. Chen, L. , 2006Real time multiple objects tracking and identification based on discrete wavelet transform, Pattern Recognition, 39 3June 2006) 1126 1139

6. J. Chiang, S. , H. Lin, T. , C. Hsia, H. , 2008Novel fast block motion estimation using diamond-arc-hexagon search patterns, Journal of the Chinese Institute of Engineers, 31 6September 2008) 955 966

7. R. T. Collins, A. J. Lipton, T. Kanade, H. Fujiyoshi, D. Duggins, Y. Tsin, D. Tolliver, N. Enomoto, O. Hasegawa, P. Burt, L. Wixson, 2000A system for video surveillance and monitoring, Carnegie Mellon University, Technical Report, (2000) CMU-RI-TR-00 12

8. S. Cvetkovic, P. Bakker, J. Schirris, P. H. N. With, de , 2006Background estimation and adaptation model with light-change removal for heavily down-sampled video surveillance signals, IEEE International Conference on Image Processing, (October 2006) 1829 1832

9. I. Daubechies, W. Sweldens, 1998Factoring wavelet transforms into lifting steps, The Journal of Fourier Analysis and Applications, 4 3 247 269

10. W. Ge, L. Gao, Q. , Q. Sun, 2007A method of multi-scale edge detection based on lifting scheme and fusion rule, International Conference on Wavelet Analysis and Pattern Recognition, 2November 2007) 952 955

11. R. C. Gonzalez, R. E. Woods, 2001 Digital image processing, Addison-Wesley Longman Publish Co., Inc., Boston.

12. C. Hsia, H. , J. Guo, M. , J. Chiang, S. , 2009An improved low complexity algorithm for 2-D integer lifting-based discrete wavelet transform using symmetric mask-based scheme, IEEE Transactions on Circuits and Systems for Video Technology, 19 8August 2009) 1201 1208

13. C. Hsia, H. , J. Guo, M. , J. Chiang, S. , C. Lin, H. , 2009A novel fast algorithm based on SMDWT for image applications, IEEE International Symposium on Circuits and Systems, (May 2009) 762 765

14. C. Hsieh, C. , S. Hsu, S. , 2007A simple and fast surveillance system for human tracking and behavior analysis, IEEE Conference on Signal-Image Technologies and Internet-Based System, (December 2007) 812 828

15. W. Hu, M. , X. Zhou, M. Hu, S. Maybank, 2009Occlusion reasoning for tracking multiple people, IEEE Transactions on Circuits and Systems for Video Technology, 19 1January 2009) 114 121

16. W. Hu, M. , T. Tan, N. , W. L. , S. Maybank, 2004A survey on visual surveillance of object motion and behaviors, IEEE Transactions on Systems, Man, and Cybernetics- Part C: Applications and Reviews, 34 3August 2004) 334 352

17. K. Huang, Q. , L. Wang, S. , T. Tan, I. , S. Maybank, 2008A real-time objects detecting and tracking system for outdoor night surveillance, Pattern Recognition, 41 1January 2008) 423 444

18. J. Huang, C. , T. Su, S. , L. Wang, J. , W. Hsieh, S. , 2004Double-change-detection method for wavelet-based moving-object segmentation, IET Electronics Letters, 40 13June 2004) 798 799

19. N. Jacobs, R. Pless, 2008Time scales in video surveillance, IEEE Transactions on Circuits and Systems for Video Technology, 18 8August 2008) 1106 1113

20. G. K. Kharate, V. H. Patil, N. L. Bhale, 2007Selection of mother wavelet for image compression on basis of nature of image, Journal of Multimedia, 2 6November 2007) 44 51

21. H. Liu, H. , X. Chen, H. , Y. Chen, G. , C. Xie, S. , 2006Double change detection method for moving-object segmentation based on clustering, IEEE International Symposium on Circuits and Systems, (May 2006) 5027 5030

22. S. G. Mallat, 1989A theory for multi-resolution signal decomposition: The wavelet representation, IEEE Transaction on Pattern Analysis and Machine Intelligence, 11 7July 1989) 674 693

23. S. J. Mckenna, 2000Tracking groups of people, Computer Vision and Image Understanding, 80 1October 2000) 42 56

24. B. Sugandi, H. Kim, J. K. Tan, S. Ishikawa, 2007Tracking of moving objects by using a low resolution image, International Conference on Innovative Computing, Information and Control, (September 2007) 408 408

25. F. A. Tab, G. Naghdy, A. Mertins, 2007Multiresolution video object extraction fitted to scalable wavelet-based object coding, IET Image Processing, 1 1March 2007) 21 38

Chapter 2

Wavelet diagnosis of ECG signals with kaiser based noise diminution

Sridhathan Chandramouleeswaran[1], Ahmed M. A. Haidar[2], Fahmi Samsuri[1]

[1]Faculty of Electrical and Electronics Engineering, University Malaysia Pahang, Kuantan, Malaysia

[2]School of Electrical, Computer and Telecommunications Engineering, University of Wollongong, New South Wales, Australia

ABSTRACT

The evaluation of distortion diagnosis using Wavelet function for Electrocardiogram (ECG), Electroencephalogram (EEG) and Phonocardiography (PCG) is not novel. However, some of the technological and economic issues remain challenging. The work in this paper is focusing on the reduction of the noise interferences and analyzes different kinds of ECG signals. Furthermore, a physiological monitoring system with a programming model for the filtration of ECG is presented. Kaiser based Finite Impulse Response (FIR) filter is used for noise reduction and identification of R peaks based on Peak Detection Algorithm (PDA). Two approaches are implemented for detecting the R peaks; Amplitude Threshold Value (ATV) and Peak Prediction Technique (PPT). Daubechies wavelet transform is applied to analyze the ECG of driver under stress, arrhythmia and sudden cardiac arrest signals. From the obtained results, it was found that the PPT is an effective and efficient technique in detecting the R peaks compared to ATV.

KEYWORDS

Diagnostic Distortion; Wavelet Transforms; ECG; Noise Interferences, FIR Filter; Kaiser Window

1. INTRODUCTION

Nowadays, Computer aided ECG signal analysis has gained thrust and incredible amount of work were carried out using these technologies for heart diagnosis. The ECG signal is characterized by five peaks and valleys labeled by the letters P, Q, R, S, T. In some cases, another peak called U is also used. The P-wave represents the activation of the upper chambers of the heart, atria, while the QRS complex and T-wave represent the excitation of the ventricles or the lower chamber of the heart [1, 2]. Normal ECG signal and its labeling are depicted in **Figure 1** [3]. ECG signal contains anatomic and physicologic aspects of the whole cardiac muscle and other information that can be exploited in different manners.

ECG analyzing system depends mainly on the accurate and reliable detection of the QRS complex, T and P waves [1,2]. Electrodes and cables are connected to the patient in examination room and monitoring is carried out in other location. Hence, electromagnetic interference of the power frequency and some other biomedical signals like baseline wander, Electromyogram (EMG) and EEG are added to the ECG signal. ECG signal frequency is approximately between 0.5 Hz and 100 Hz. Baseline Wander frequency is below 1 Hz and EEG signal frequencies are above 100 Hz. EMG frequency can overlap with ECG frequency depending on body muscle movement which can be reduced to the barest by making the patient staying still and quiet [4-6].

The wavelet transform is a time-scale representation method that decomposes signals into basis functions of time and scale, which makes it useful in many applications such as signal de-noising, wave detection, data compression and feature extraction [1,7].

The wavelet transformation is based on a set of analyzing wavelets allowing the decomposition of ECG signal in a set of coefficients. Each analyzing wavelet has its own time duration, time location and frequency band [1]. In the wavelet based algorithm, the ECG signal is de-noised by removing the corresponding wavelet coefficients at higher scales [2]. One of the main applications of a filter is to sort out the unwanted parts of an

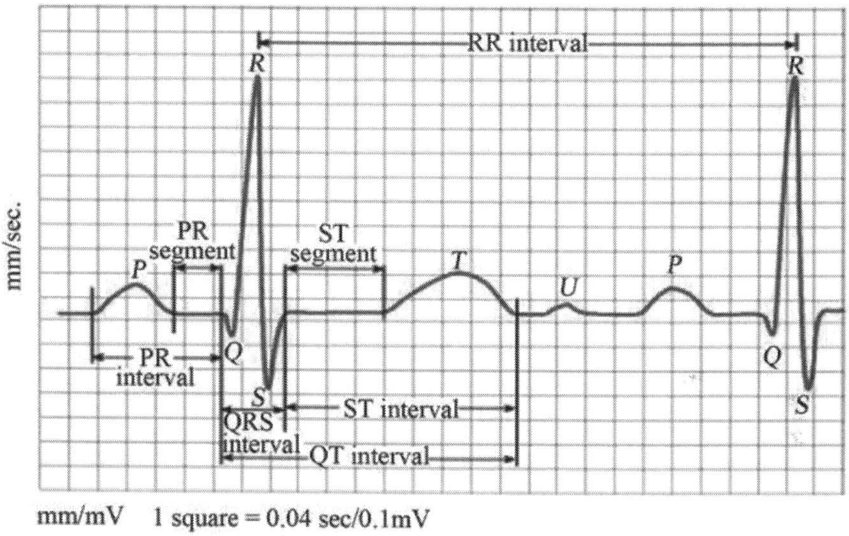

Figure 1. Normal ECG Signal with labeling.

input signal. These unwanted parts might be noise present out of the frequency band in the desired signal. During transforming the corrupted signal into the frequency domain, the frequencies that belong to the external noise can be easily detected [7,8].

Wavelet transform are used to compress the ECG signal. The main challenge in an ECG compression method is to minimize the storage requirements without losing the clinically significant information which can be achieved using Wavelet transform. Small variations in normal and noise corrupted ECG signal can been extracted using wavelet function [9,10].

The wavelet based signal de-noising is performed using wavelet shrinkage and thresholding technique. There are two types of thresholdings; hard and soft. In hard threshold, the elements whose absolute value is lower than the threshold are set to zero. Soft threshold is an extension of hard threshold. Here, the elements whose absolute values are lower than the threshold are first settled to zero, and then shrinking the nonzero coefficients towards 0 [11].

The series connection of Discrete Wavelet Transform, thresholding and Inverse Discrete Wavelet Transform can remove the noises and achieve high Signal-to-Noise ratios. This is due to the concentrating ability of the wavelet transform [12]. The combination of Shannon energy envelope estimator and Hilbert-transform technique can detect R-peaks in case of wider and small QRS complexes, negative QRS polarities, and sudden changes in QRS amplitudes [13].

Sharing the patient information and exchange of medical images and Electronics Patient Record between networked hospitals and healthcare centers is witnessed nowadays. Discrete wavelet packet transforms is used to protect the patient information in the form of watermark on the medical image using the hospital logo as a reference image. Since wavelet based watermarking method is robust against a wide range of attacks such as JPEG image format, contrast adjustment, Gaussian noise, histogram equalization, gamma correction etc. [14].

ECG analysis can be done reliably and accurately if the noises and artifacts are removed. The noise reduction in ECG has been reported in many research works. In our work the filtration of ECG is done using Kaiser based FIR filter which has attractive character both in time and frequency domain. Kaiser window has the ability to remove noise effectively than other windowing techniques. Here FIR filters are designed to remove the noise in a particular frequency range such as 0.5 Hz to remove the Baseline Wander, 50 Hz to remove the power line interference etc. Filtration is done prior to R peak detection to minimize the false detection rate. It can be seen from the results that the designed filter removes noise effectively. In many previous works the filtration is done on normal ECG signal. Wavelet transform is applied on normal or abnormal ECG such as tachycardia, bradycarida. In our work, we apply wavelet analysis for signals such as sudden cardiac arrest, and driver in stress.

The aim of the work is to remove the noises from the ECG signals and identify the R peak. After the noise removal, the ECG peaks are detected by the Peak Detection Algorithm (PDA) techniques. The wavelet of the abnormal ECG signals is compared with the normal one. The data for this analysis was collected from PhysioBank data base [15]. Section 1 gives a brief introduction about the ECG signals. Section 2 outlines the function of Kaiser based noise reduction using FIR filters. Wavelet transform is described in Section 3. Approach to PDA and Wavelet analysis is described in Section 4. Section 5 presents the results and discussion followed by conclusion.

2. KAISER BASED NOISE REDUCTION USING FIR FILTERS

FIR filter is one of the techniques successfully employed in processing ECG signals for measurement and noise reduction. They are used to modify the bio-signal by removing noise and the effect of a system on the input signal can be described in both time and frequency domains. The time domain method of ECG signal analysis is not always sufficient to study all the features of ECG signals. Therefore, the frequency representation of a signal is required [5,6]. Hence, filters are designed in frequency domain so that the noisy signals can be filtered from the measured bio-signals. If this filters where designed in time domain it would not be possible to remove the noises effectively since the frequencies that belong to the external noise cannot be detected easily. In this section more details about the filters and filtration process are presented.

2.1. Notch Filter (NF)

The noise presented at 50/60 Hz is known as a power line hum (noise), caused by any electrical appliances and lights [5,6]. NF is used to filter this noise. The frequency response of NF is shown in **Figure 2** and the transfer function of a NF is given by,

$$H_{\text{notch}}(Z)=\frac{b_o\left(z-e^{j\theta_0}\right)\left(z-e^{-j\theta_0}\right)}{\left(z-re^{j\theta_0}\right)\left(z-re^{-j\theta_0}\right)} \tag{1}$$

where,

$$r \approx 1-\frac{\Delta F\pi}{f_a},\ b_0=\frac{\left|1-2r\cos(\theta_0)+r^2\right|}{2\left|1-\cos(\theta_0)\right|} \tag{2}$$

B_0 = filter coefficient,

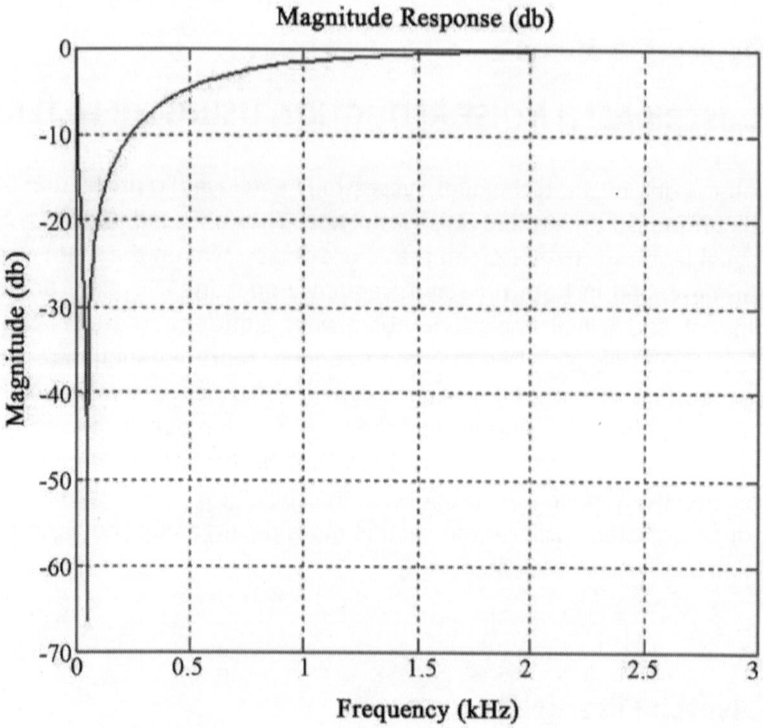

Figure 2. Frequency response of a 50 Hz NF.

z = unit delay operatorr = pole radius,

$\theta_0 = 2\pi F_0 T$ F_0 = Notch frequencyfs = sampling frequency,

ΔF = Notch Band.

2.2. High Pass Filter (HPF)

The HPF removes the corrupting low frequency noises in ECG signal. HPF is designed with the cut off frequency is 0.5 Hz and the sampling frequency is 6 KHz. The frequency response of designed HPF is shown in **Figure 3**. The transfer function of an ideal HPF is determined by the following constraints,

$$H_{HP}(\omega) = \begin{cases} 0 & |\omega| < \omega_c \\ 1 & |\omega| > \omega_c \end{cases} \quad (3)$$

where $\omega_c = 2\pi f_c$ is the filter cut-off frequency.

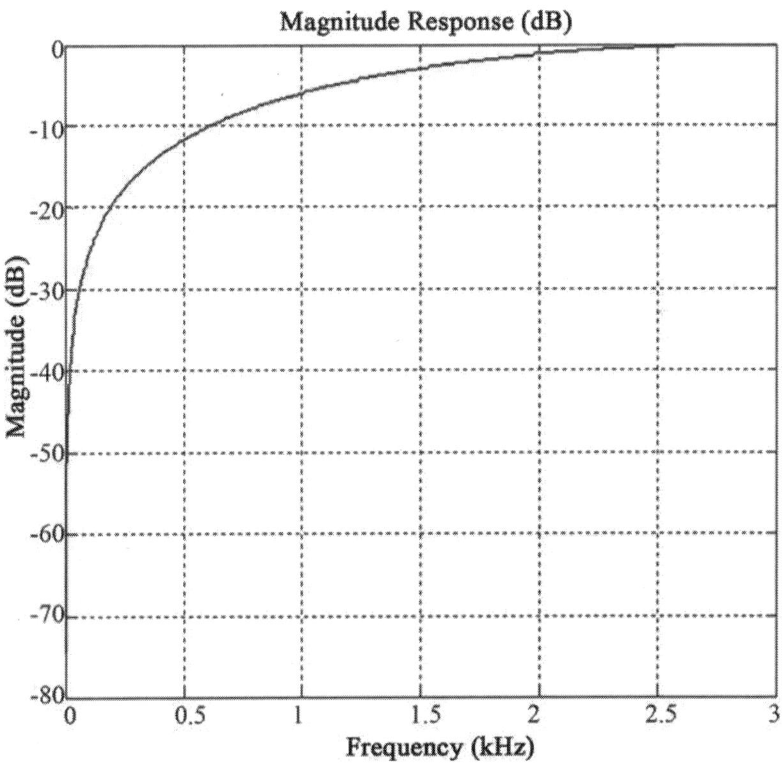

Figure 3. Frequency response of HPF.

2.3. Low Pass Filter (LPF)

The LPF removes the corrupting high frequency noises in ECG. The transfer function of an ideal LPF is given by Equation (4). Here the LPF is designed with the cut off frequency (f_c) 100 Hz with the sampling frequency (fs) of 6 KHz. The frequency response of designed low pass filter is shown in**Figure 4**.

$$H_{LP}(\omega) = \begin{cases} 0 & |\omega| < \omega_c \\ 1 & |\omega| > \omega_c \end{cases} \quad (4)$$

2.4. Kaiser Window (KW)

Kaiser based FIR filter is designed to remove the unwanted frequency component which corresponds to the noise. Kaiser window has very desirable characteristics both in time domain and frequency domain. A typical KW function is depicted in **Figure 5**. A good window should be a time limited function with a Fourier transform that is band limited and Kaiser Window possesses such characteristics [6]. The Kaiser window is given by,

$$\omega_k(n) = \begin{pmatrix} \dfrac{I_0\left[\pi\alpha\sqrt{1-\left[\dfrac{2n}{N-1}\right]^2}\right]}{I_0(\pi\alpha)} & \text{for} \lfloor n \rfloor \leq \dfrac{N-1}{2} \\ 0 & \text{otherwise} \end{pmatrix} \quad (5)$$

where α is the adjustable parameter and $I_0(x)$is the modified zero order Bessel functions of the first kind. In some works the factor β is also used, in this case Equation (5) is modified into,

$$\omega_k(\beta, n) = \begin{pmatrix} \dfrac{I_0\left[\beta\left\{1-\left[\dfrac{2n}{N-1}\right]^2\right\}^{\frac{1}{2}}\right]}{I_0(\pi\alpha)} & \text{for} \lfloor n \rfloor \leq \dfrac{N-1}{2} \\ 0 & \text{otherwise} \end{pmatrix} \quad (6)$$

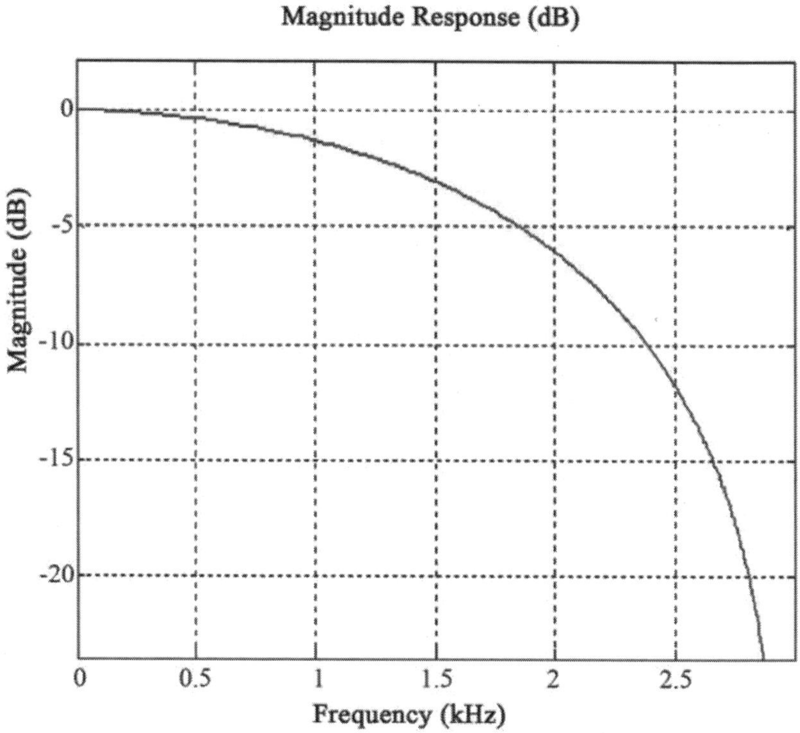

Figure 4. Frequency response of LPF.

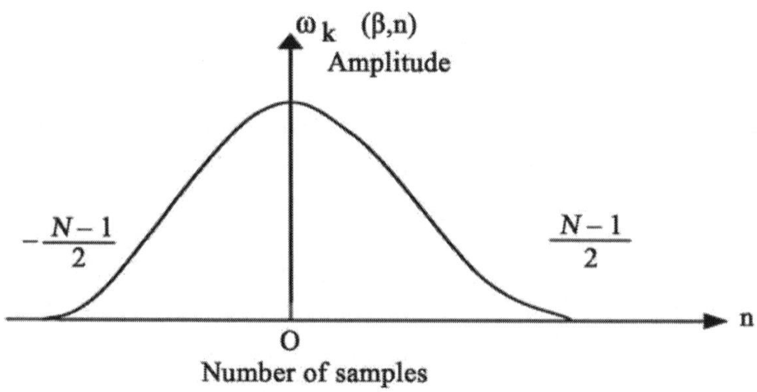

Figure 5. Kaiser window function.

3. WAVELET TRANSFORM

A signal as the function of $f(t)$ can often be better analyzed and expressed as a linear decomposition of the sums; coefficient products and function. In the wavelet transform, the original signal is transformed using predefined wavelets. These wavelets are orthogonal, orthonormal, biorthogonal, scalar or multi wavelets. The wavelet transform is a convolution of the wavelet function $\psi_{m,n}(t)$ with thesignal $x(t)$. Orthonormal dyadic discrete wavelets are associated with scaling functions $\varphi(t)$. The scaling function can be convolved with the signal to produce approximation coefficients "S" [1,2,8, 10-12,16]. The Discrete Wavelet Transforms (DWT) can be written as,

$$T_{m,n} = \int_{-\infty}^{\infty} x(t)\psi_{m,n}(t)\,\mathrm{d}t \tag{7}$$

The original signal can be reconstructed by choosing orthonormal wavelet with the approximation coefficient as,

$$S_{m,n} = \int_{-\infty}^{\infty} x(t)\phi_{m,n}(t)\,\mathrm{d}t \tag{8}$$

The signal approximation at a specific scale is a combination of the approximation and detail at the next lower scale is,

$$x_m(t) = x_{m-1}(t) - d_m(t) \tag{9}$$

In order to keep the total number of data points similar at each level, every other data point of the output sequences is discarded; this process is known as downsampling. Using up-sampling the original data set is reconstructed [17].

In discrete, the wavelet transform is modified to a filter bank tree using the decomposition/reconstruction as shown in Figures 6 and 7. ECG signal is decomposed into a sequence of wavelet coefficients using linear phase bi-orthogonal filter bank DWT.

Referring to **Figure 6**, the input signal is fed to "h" and "g", where "h" is the high pass filter and "g" is a low pass filter. The output of the high pass filter is down sampled to get detailed coefficients and approximation coefficients from low pass filter after down sampling. If there are many levels of decompositions, at each level, the high pass filter produces detail information, while the low pass filter produces approximation information.

As seen from **Figure 7**, the approximation and detail coefficients at every level are up-sampled, passed through the low pass and high pass synthesis filters and then added. Same number of levels as in the decomposition process must be used to obtain the original signal.

The Decomposition Coefficient Vector (DCV) is obtained by concatenating the Approximation band Coefficients (AC5) at the fifth level and the entire Detail band Coefficients (DC5 - 1) start from the last level of the decomposition to first level [10-12,16]. The decomposition coefficient vector is given by,

$$DCV = AC5 + DC5 + DC4 + DC3 + DC2 + DC1 \tag{10}$$

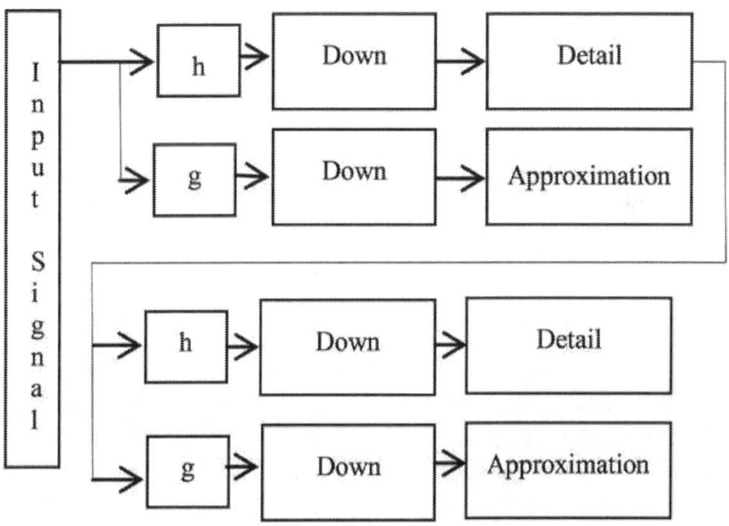

Figure 6. Filter bank tree decomposition.

In this work Daubechies [1,3] wavelet analysis is performed on ECG signals of automobile drivers under stress and sudden cardiac arrest. Although the Daubechies algorithm is conceptually more complex and has a slightly complicated computation, yet this algorithm picks up minute detail that is missed by other wavelet algorithms, like Haar wavelet algorithm. For Daubechies wavelet transforms the decomposition is given by,

$$c(n) = h_0x(2n)+h_1x(2n+1)+h_2x(2n+2)+h_3x(2n+3) \quad (11)$$

$$d(n) = h_3x(2n)-h_2x(2n+1)+h_1x(2n+2)-h_0x(2n+3) \quad (12)$$

where the multipliers are found by,

$$h_0=\frac{1+\sqrt{3}}{4\sqrt{2}},h_1=\frac{3+\sqrt{3}}{4\sqrt{2}},h_2=\frac{3-\sqrt{3}}{4\sqrt{2}},h_3=\frac{1-\sqrt{3}}{4\sqrt{2}} \quad (13)$$

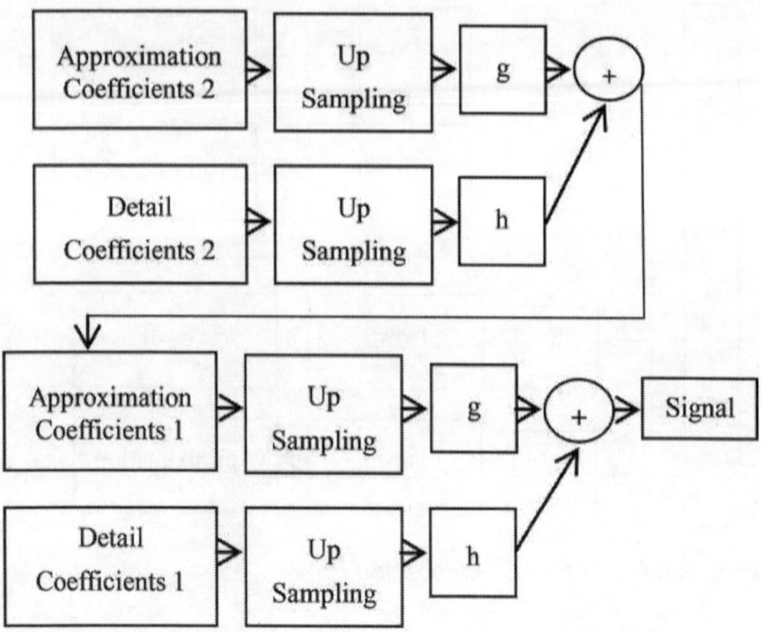

Figure 7. Filter bank tree reconstruction.

4. PEAK DETECTION ALGORITHM IN WAVELET ANALYSIS

Biomedical signals are needed to be filtered so that the physician can evaluate the patient condition clearly. The QRS complex in the ECG reflects the electrical activity of heart during the ventricular contraction, the time of its occurrence and shape provide good information about the current state and condition of the heart based on peak detection for QRS signals using threshold values. In this work, FIR filter with Kaiser Window approach was selected because of its ability to remove noise in the ECG. The noise corrupted ECG signal is applied to the 50 Hz NF in order to remove the power line interference or hum. Deviations in the normal ECG patterns indicate various cardiac disorders.**Table 1** demonstrates features of various abnormalities and their characteristic [1].

The proposed approach is used to remove the power line interference, baseline wander frequency and other noisy signals which overlap with ECG frequency and also to identify the R peaks in ECG. In this approach, Filtered ECG signal is fed to the PDA using two approaches ATV and PPT so that the R peaks can be detected. After filtration, signal is scanned for R peaks using PDA. The flowchart of ATV PDA algorithm is shown in **Figure 8** and the steps of such approach are listed.

1) ECG signals whose values are positive are considered and the negative values are neglected using zero threshold ($x(s) \geq 0$), retain same value, else zero;

2) Amplitude Threshold Value (ATV) is calculated from the ECG signal (ATV = 0.7 maximum amplitude value);

3) ATV is used to rectify the ECG signal ($x(s) \geq$ ATV), retain same value, else zero;

4) In ECG signals, R peaks are detected by clipping the unwanted parts by using the ECG clipper. Clipper uses logical conditions to identify the R peaks;

5) The detected peaks are displayed.

Table 1. Various abnormalities and their characteristic features.

S. No	Name of abnormality	Characteristic features
1	Dextrocardia	Inverted P-wave
2	Tachycardia	R-R interval <0.6 s
3	Bradycardia	R-R interval >1 s
4	Hyperkalemia	Tall T-wave and absence of P-wave
5	Myocardial ischaemia	Inverted T-wave
6	Hypercalcaemia	QRS interval <0.1 s
7	Sinoatrial block	Complete drop out of a cardiac cycle
8	Sudden cardiac arrest	Irregular ECG

The flowchart of PPT is depicted in **Figure 9**. As seen from flowchart, PPT identifies the highest value point between valleys. This method searches the maximum peak value in the signal and stores the value and the address location. It compares stored values to define the peak point in the signal.

The performances of the PDA can be assessed by evaluating the sensitivity, positive prediction and false negatives as,

$$\text{Sensitivity} = \frac{TP}{TP+FN} \quad (14)$$

$$\text{Positive Prediction} = \frac{TP}{TP+FP} \quad (15)$$

$$\text{Error rate} = \frac{FN+FP}{\text{No of QRS}} \quad (16)$$

whereTP: number of true positive detections;

FN: number of false negatives;

FP: number of false positives.

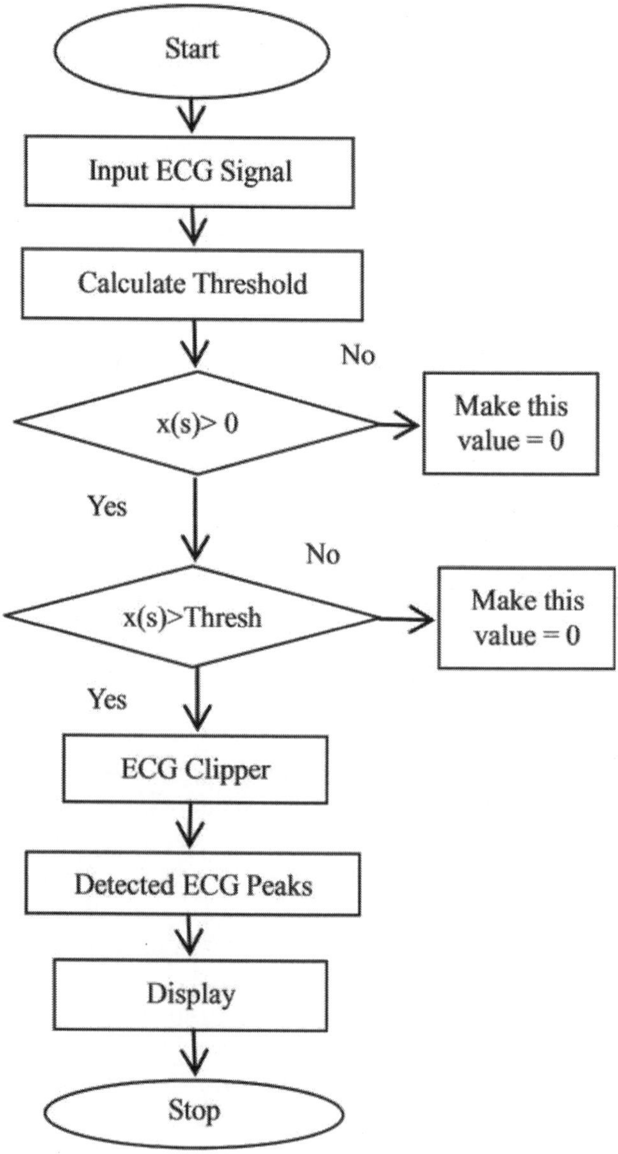

Figure 8. The flowchart of ATV PDA.

5. RESULT AND DISCUSSION

The crucial step in the ECG analysis is to remove the noise and differentiate waves forming the entire cardiac cycle. Matlab environment was used to design and evaluate the performance of the filter in removing noise and to analysis the wavelet of different ECG signals. The design uses FIR filter with Kaiser Window built in the Matlab. The NF designed here is used to remove the 50/60 Hz power line interference. FIR filter designed for the pass band and stop band frequency of 0.5 Hz and 100 Hz with Kaiser Window. In Figures 10-12, (a) is the ECG signal before filtering is displayed in green colour and ECG signal after filtering is shown in blue colour and (b) the noise removed from ECG. It can be observed that the filter reduced noises in ECG to a considerable level as seen in Figures 10-12. This filtered ECG signal is passed through PDA techniques.

Figures 13-16 show the output after applying the PDA to the ECG signals. In these figures, the first part is the ECG signal whose peak has to be detected. The second part is the detected peaks which are incorporated and highlighted in ECG signals. Third part is the peaks that are detected. Figures 13-14 show the results after applying ATV to the ECG signals. As seen from these figures, ATV detects the R peaks when the noise level is less. If noise level or artifacts are more in ECG, the ATV fails to detect the R Peaks. Figures 15-16 show the results after applying PPT to the ECG signals. From these figures it is clearly seen that PPT detects the R peaks when the noise level or the artifacts in ECG is more. The false detection is also less in PPT compared to ATV. From information given in**Table 2**, the sensitivity, positive prediction and error rate can be calculated.

Daubechies wavelet transform is applied using Wavelet GUI in Matlab on the ECG signals such as arrhythmia, drivers under stress and sudden cardiac arrest. Here five levels of dB wavelet decomposition (d1 to d5) and approximations (a1 to a5) are performed on the ECG sig nals. The steps involved for applying wavelet transform to the signals using MATLAB tools is described here. The wavelet GUI is opened using command window by typing wave menu, wavelet 1-D is selected from the menu. The signal is imported and required wavelet transform is applied on the signals and based on that, the decomposition and approximation coefficients are obtained and analyzed. The Daubechies wavelet transform is applied for the different ECG signals taken for physiobank database and the results of these analyses are graphically shown in Figures 17-20.

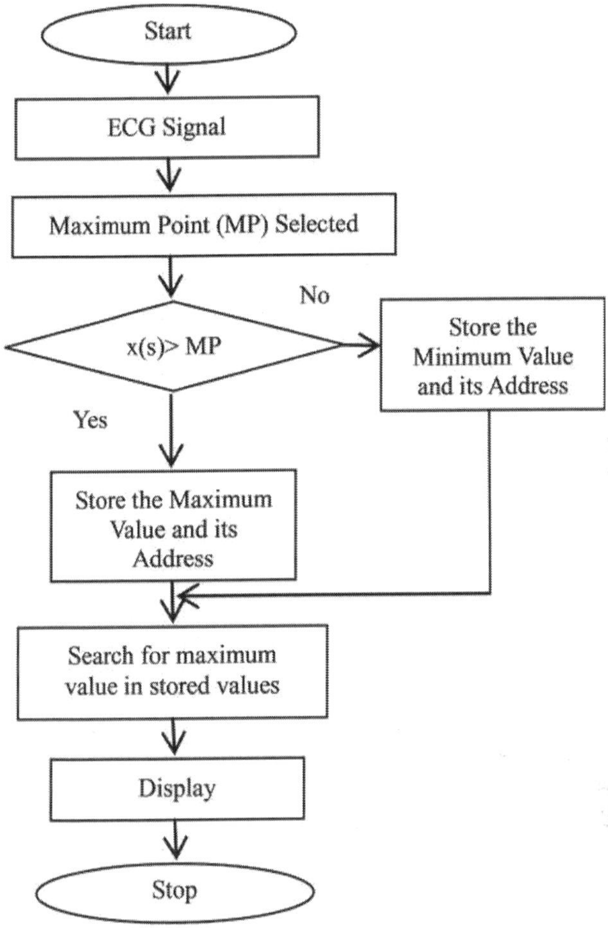

Figure 9. The flowchart of PPT.

The wavelet transformation is worth investigating especially in P and T wave recognitions. The peaks of QRS complexes flatten; P and T waves containing lower frequencies become more visible [2]. In**Figure 17**(a) (left hand side) "S" is the normal ECG wave and a1 to a5 are the approximation coefficients. ECG signal, d1 to d5 decomposition coefficients are shown on the right hand side of the**Figure 17**. It can clearly be seen from the decomposition coefficients d1 to d3, that the frequency and amplitude components correspond directly to the QRS complex. In higher decomposition coefficients, the P and T components are more visible.

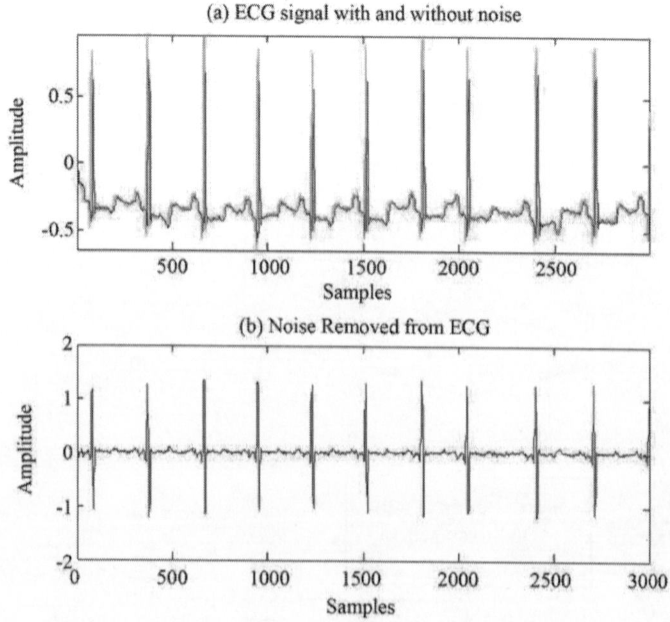

Figure 10. ECG signal before and after filtration (green color ECG signal with noise, blue color ECG signal after filtration).

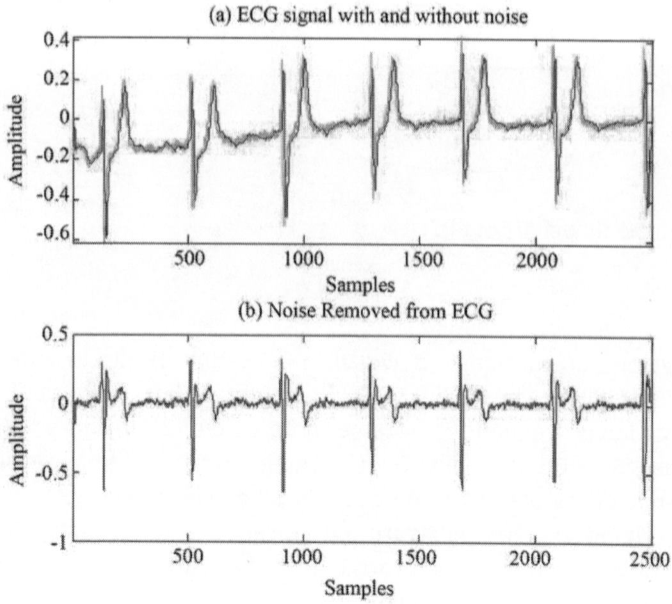

Figure 11. ECG signal before and after filtration.

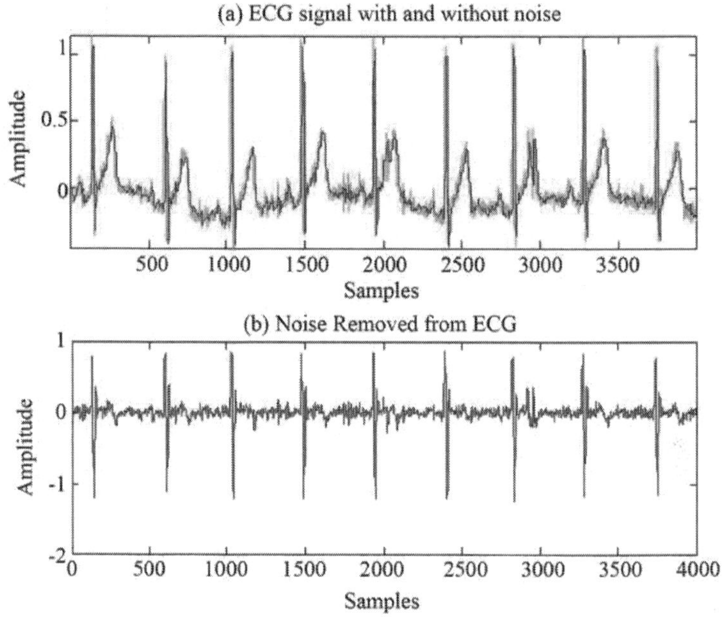

Figure 12. ECG signal before and after filtration.

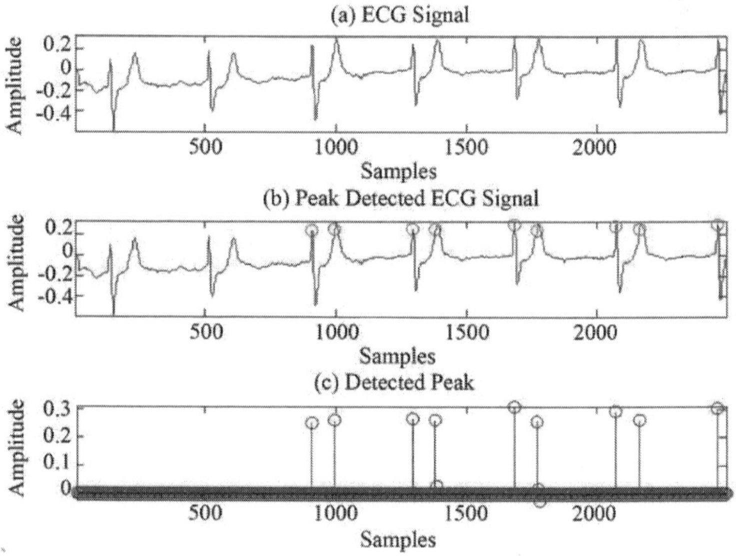

Figure 13. R peaks detection using ATV.

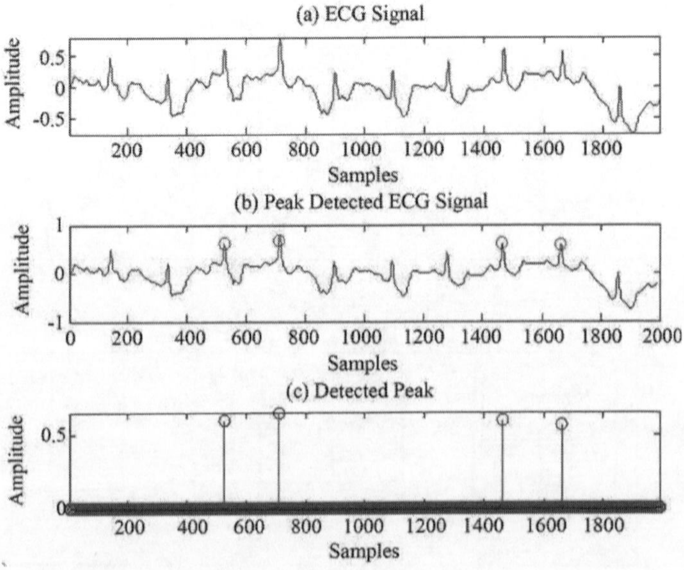

Figure 14. R peaks detection using ATV.

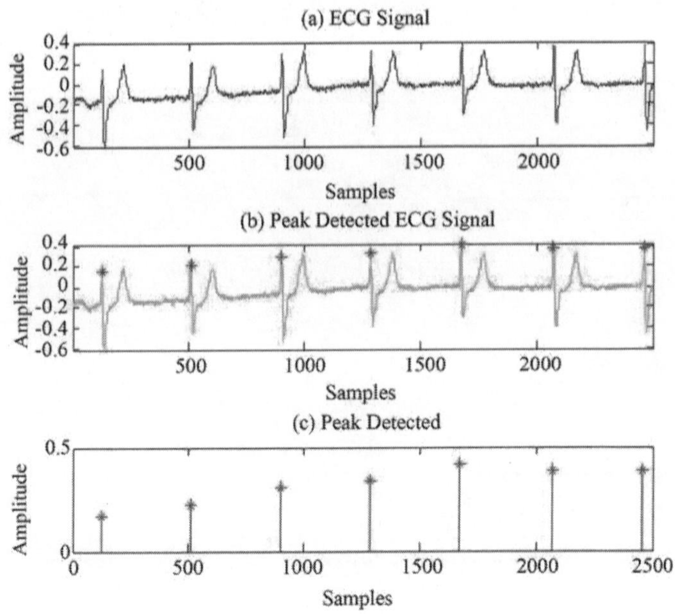

Figure 15. R peaks detection using PPT.

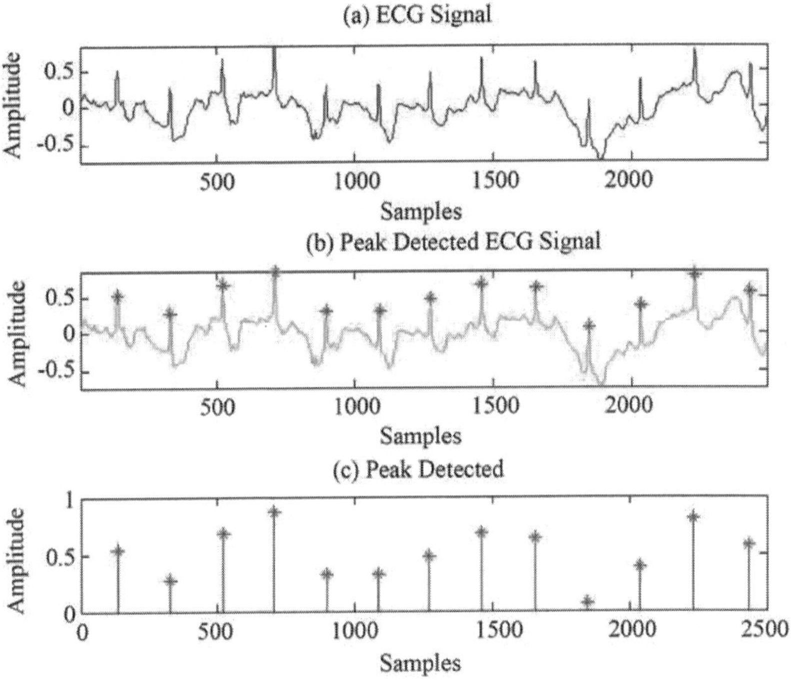

Figure 16. R peaks detection using PPT.

Table 2. Performance characteristic features of PDA.

PDA method	Sample	No. of QRS	True positive	False negative	False positive
Amplitude threshold value	1	7	5	2	4
Amplitude threshold value	2	10	4	6	-
Peak prediction technique	1	7	7	-	-
Peak prediction technique	2	13	13	-	-

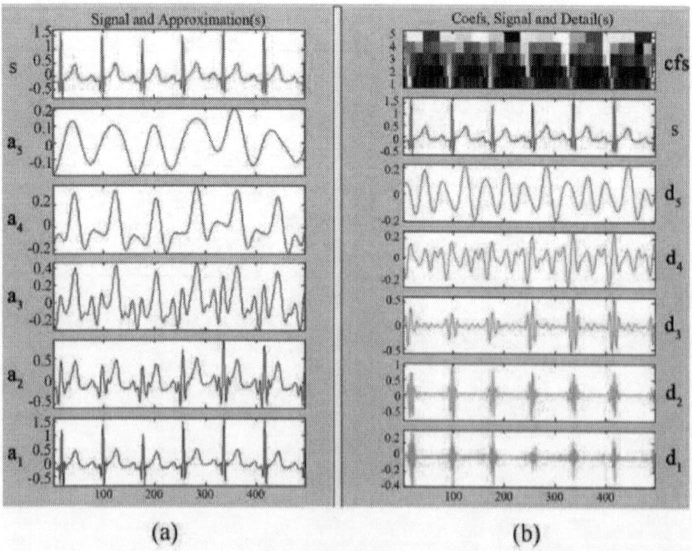

Figure 17. (a) ECG and corresponding approximate coefficients; (b) Decomposition coefficients.

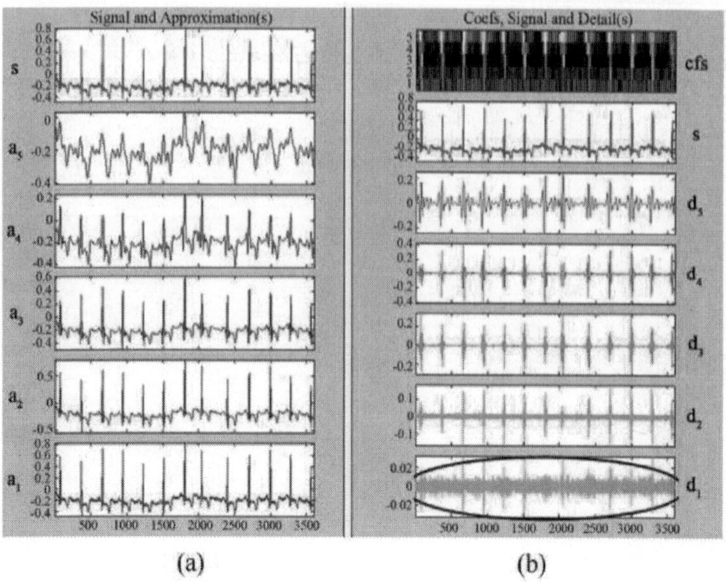

Figure 18. (a) Arrhythmia Signal and corresponding approximate coefficient; (b) Decomposition coefficients.

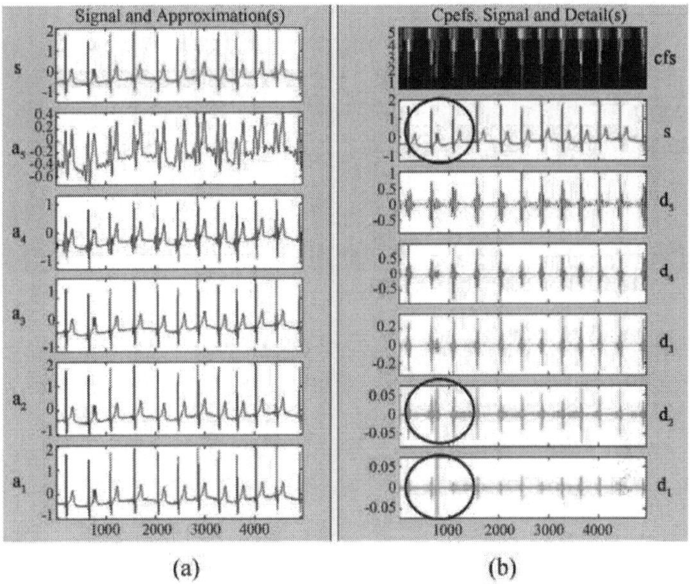

Figure 19. (a) ECG of driver under stress and corresponding approximate coefficients; (b) ECG of decomposition coefficients.

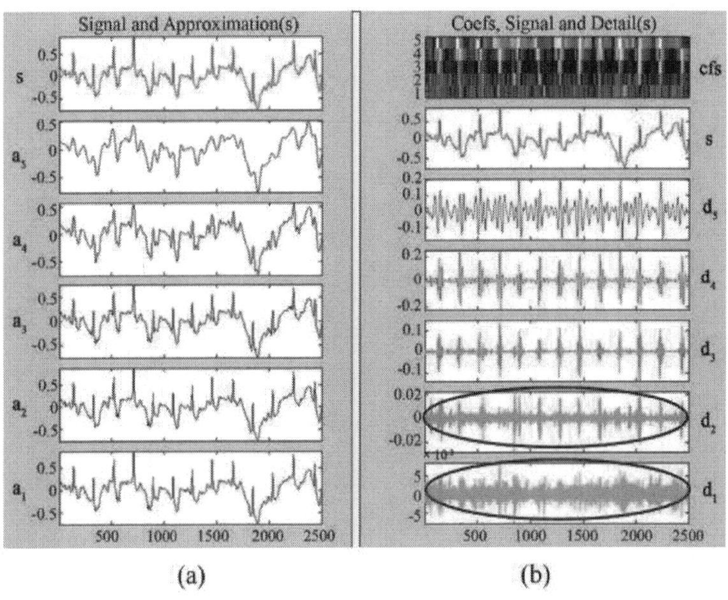

Figure 20. (a) Sudden cardiac arrest and corresponding approximate coefficients; (b) Decomposition coefficient.

In **Figure 18**, arrhythmia signal with approximation coefficients a1 to a5 are shown on the left hand side. Signal and its detailed coefficients d1 to d5 are shown on right hand side of **Figure 18**. In this signal, the duration of each cardiac cycle is not regular, it can be seen in the 7th and 8th cycle T and P wave are merged and in 8th cycle T wave is larger than the normal duration. When arrhythmia d1 coefficients is compared with the normal signal, it can be noted that d1is having more change in frequency which is highlighted by oval in **Figure 18**. The d4 and d5 coefficients have high frequency variation compared to normal signal.

ECG Signals of an automobile driver under stress with approximation coefficient a1 to a5 and detailed coefficients d1 to d5 are shown in the left and right hand side of **Figure 19** respectively. When the driver is in stress the ECG wave is having drastic change and deviates from the normal wave. From **Figure 19**, it is seen that the P wave is missing in the entire signal. In the second cardiac cycle, T wave is repeated that leads to amplitude and frequency shift in detailed coefficients d1, d2 which is highlighted by circle in **Figure 19**(b).

ECG signal of sudden cardiac arrest, its five scales of approximation coefficient a1 to a5 and decomposition coefficients d1 to d5 are shown in the left and right hand side of **Figure 20**. It is noted that the ECG is irregular from the normal wave pattern. In d1 and d2 coefficients, there are more changes in frequency compared to the normal ECG wave as highlighted. The decomposition coefficient d5 is extensively spread in the entire range of the signal with varying frequencies and amplitudes. From the results it can be concluded that detailed coefficients of wavelet transform can be used to analyses the normal and abnormal ECG signal.

6. SUMMARY

The objective of this work is ECG noise filtration, R Peak detection and wavelet analysis. The ECG data's are obtained from the physio-bank database. From the database the ECG signals considered are normal, ECG of driver under stress, Sudden Cardiac arrest and Arrhythmia. FIR filter with Kaiser Window is used to remove the noises. The R-peak is detected in two techniques: a) Amplitude Threshold Value; b) Peak Prediction Technique. ATV uses threshold value to detect the peaks and is not effective and fails when ECG is having more artifacts. PPT identifies the R-peaks by combining threshold value and peak address position. So the R-peaks are detected more effectively compared to ATV. Analysis of ECG is done using Daubechies wavelet transform. The decomposition coefficients are considered and compared here. The d1 to d5 coefficients are studied for normal, ECG of driver under stress, Sudden Cardiac arrest and Arrhythmia. It was found that d1 to d3 coefficients reflects more in phase and frequency change in the abnormal signals compared to normal. PPT and wavelet is very useful tool for analyzing the ECG signals and help the physicians to identify cardiac disorders.

7. CONCLUSION

In this paper, ECG analysis using wavelet transform with PDA is presented. Kaiser Window incorporated with FIR filter is proposed as a tool for noise diminution. This approach has been tested for normal and abnormal ECG signals collected from the physio-bank database. Wavelet diagnosis helps in identifying the abnormalities in the ECG signal and it is reflected in its detailed coefficients. It can be concluded that the designed Kaiser based filter can remove the noises efficiently from the ECG. PPT is capable to detect the R-peak in ECG signal effectively when compared to ATV. The wavelets transform can be used in processing of ECG and other non-stationary signals. Most of the features in ECG signals can be obtained through the application of the wavelet transform. If the wavelet coefficients are having any amplitude or frequency variations compared with the normal ECG, it means that the ECG signal is abnormal. As know from the medical diagnosis, the ECG is extensively used for heart diseases and the high quality of signal is necessary to be achieved. Therefore, the presented technique in this work will help the physicians to identify clearly the deviations from the regular ECG patterns that indicate various cardiac disorders.

REFERENCES

1. Saritha, C., Sukanya, V. and Murthy, Y.N. (2008) ECG signal analysis using wavelet transforms. Journal of Chemical Physics, 35, 68-77.

2. Narayana, K.V.L. and Rao, A.B. (2001) Wavelet based QRS detection in ECG using MATLAB. Innovative Systems Design and Engineering, 2, 2222-1727.

3. Bakul, G. and Tiwary, U.S. (2010) Automated risk identification of myocardial infarction using relative frequency band coefficient (RFBC) features from ECG. The Open Biomedical Engineering Journal, 4, 217-222.

4. Luo, S. and Johnston, P. (2010) A review of electrocardiogram filtering. Journal of Electrocardiology, 43, 486-496.

5. Lee, J.-W. and Lee, G.-K. (2005) Design of an adaptive filter with a dynamic structure for ECG signal processing. International Journal of Control, Automation, and Systems, 3, 137-142.

6. Mbachu, C.B., Onoh, G.N., Idigo, V.E., Ifeagwu, E.N. and Nnebe, S.U. (2011) Processing ECG signal with Kaiser Window-based FIR digital filters. International Journal of Engineering Science and Technology, 3, 6775-6783.

7. Mahesh, S.C., Agarwala, R.A. and Uplane, M.D. (2008) Suppression of baseline wander and power line interference in ECG using digital IIR filter. International Journal of Circuits, Systems and Signal Processing, 2, 356-365.

8. Leonardo, V.B., Elmar, U.K.M. and Luis, C.C. (2001) Compression of ECG signals by optimized quantization of discrete cosine transform coefficients. Medical Engineering & Physics, 23, 127-134.

9. Manikandan, M.S. and Dandapat, S. (2006) Wavelet threshold based ECG compression using USZZQ and Huffman coding of DSM. Biomedical Signal Processing and Control, 1, 261-270.

10. Haque, K.M.F., Ali, Md.H., Kiber, M.A. and Hasan, Md.T. (2009) Detection of small variations of ECG features using Wavelet. ARPN Journal of Engineering and Applied Sciences, 4, 27-30.

11. Zoltan, G.-S. and Petru, M. (2007) Wavelet transform based ECG signal denoising. Scientific International Conference, Romania, 15-16 November 2007, 1-7.

12. Abdel-Rahman, A.-Q. and Khaled, D. (2010) ECG signal enhancement using wavelet transform. WSEAS Transactions on Biology and Biomedicine, 7, 62-72.

13. Manikandan, M.S. and Soman, K.P. (2012) A novel method for detecting R-peaks in electrocardiogram (ECG) signal. Biomedical Signal Processing and Control, 7, 118-128.

14. Salwa, A.K.M., Naser, El-S., Tolba, A.S., Abdelkader, F.M. and Hisham M.E. (2010) Wavelet packets-based blind watermarking for medical image management. The Open Biomedical Engineering Journal, 4, 93-98.

15. Goldberger, A.L., Amaral, L.A.N., Glass, L., Hausdorff, J.M., Ivanov, P.Ch., Mark, R.G., Mietus, J.E., Moody, G.B., Peng, C.K. and Stanley, H.E. PhysioBank. PhysioToolkit, and PhysioNet: Components of a New Research Resource for Complex Physiologic Signal. http://physionet.ph.biu.ac.il/physiobank/database/mitdb/

16. Manikandan, M.S. and Dandapat, S. (2007) Wavelet energy based diagnostic distortion measure for ECG. Biomedical Signal Processing and Control, 2, 80-96.

17. John, D.E. and Joseph, D.B. (2011) Introduction to biomedical Engineering. 3rd Edition, Academic Press, Waltham, 2011.

Chapter 3

Comparison between Fourier and Wavelets Transforms in Biospeckle Signals

Kleber Mariano Ribeiro[1], Roberto Alves Braga Júnior[1], Thelma Sáfadi[2], Graham Horgan[3]

[1]Engineering Department, Federal University of Lavras, Lavras, Brazil

[2]Exact Science Department, Federal University of Lavras, Lavras, Brazil

[3]Biomathematics and Statistics Scotland, Rowett Institute of Nutrition and Health, Aberdeen, Scotland

ABSTRACT

The dynamic speckle is a non-destructive optical technique that has been used as a tool for the characterization of the biological activity and several studies are conducted to obtain for more information about the correspondence of the observed phenomena and their expressions in the interference images. Analysis in the frequency domain has been considered as powerful alternative, and although there are works using Fourier transform in the frequency analysis of the biospeckle signals, the majority presents the wavelet transform as tool for spectral analysis. In turn, there are still doubts if the Fourier transform is not enough for the analysis of the biospeckle, which would enable the reduction of processing time since an operation is computationally simpler. In this context, the present study aims to compare the constituents' parts of the speckle signal according to Fourier and wavelet transforms for numerical analysis. The comparative analysis based on the absolute values of the differences technique (AVD) was carried out for performance evaluation of the Fourier and wavelet transforms, in which the speckle signals were decomposed spectrally and subsequently reconstructed with the elimination of specific frequency bands. Results showed that the wavelet transform allowed more information about signals constituents of the dynamic speckle, emphasizing its use instead of the Fourier transform, which in turn was restricted the situations in which the only interest is to know the spectral content of the data.

KEYWORDS

Spectral Analysis; Dynamic Speckle; Biological Activity

1. INTRODUCTION

When a coherent light, such as laser, illuminates a rough surface, compared the wavelength of laser, it occurs a phenomenon of optical interference with the formation of light and dark regions, called speckle [1].

After applying the dynamic surface, there is a continuous formation of new and different speckles, and these random and dynamic interference patterns is called dynamic speckle or biospeckle, if the area concerned is biological. This technique allows extracting information about the structures movement of the illuminated material, making it an interesting tool in several knowledge areas [2].

The biospeckle has been used as a technique to measure detailed extensions of pine roots [3] or even the biological activity of roots in tissue culture [4], in assessing the water activity in maize and beans seeds [5], to studies of the relationship between chlorophyll pigments present in apples and their respective biological activity [6], and several other papers.

The biological activity expressed in the context of speckle does not present a clear definition of what phenomenon is creating, however can be understood as structural and molecular motions occurring in the material analysis [4], Doppler effect, Brownian motion, variations of the refractive index [7], among others. It is a complex signal and with causes still investigated [8], which is a challenge, and at the same time, a motivation.

In this context, the use of image processing techniques and signal analysis tools can be used in the biospeckle signal to understand better this optical phenomenon.

The interference patterns analysis can use graphical methods, which generate maps indicating the spatial variability of the biological activity, or a numerical interpretation of the temporal variation of patterns formed. An alternative of the graphical and numerical classifications is a signal analysis in the time domain or in the frequency domain [5].

The analysis of biospeckle signals in the frequency domain has been an alternative for many applications, allowing the filter and images contrast, beyond search of frequency markers of phenomena that contribute to the formation of the interference patterns in time, as described by [5]. Thereby, Fourier and wavelet transforms can be a good choice to make such analysis in the frequency domain.

Several studies have been conducted using the spectral analysis in the biospeckle signal, such as [9] that used the Fourier transform to analysis bean seeds contaminated by two kinds of fungi and managed to differentiate them using the harmonics amplitude; [10] assessed damage in apples and seed germination using wavelet transform and defined frequency markers for biological phenomena, as well as [5] who studied maize and beans, and cancer isolation and others.

Although there are many papers applying spectral analysis in the biospeckle signal, the most journals use wavelet transform and there is no works evaluating if Fourier transform, which is simpler than wavelet transform. It's enough in the frequency analysis of the dynamic speckle. In this context, the present study aims to compare the Fourier and wavelet transform in the spectral analysis of biospeckle signal.

2. THEORY

2.1. Time History of the Speckle Patterns (THSP)

The biospeckle is a nondestructive optical technique based on the analysis of the variations of the laser light scattered from material, and the biological activity presented reflects the state of the investigated object [11].

Follow a set of pixels of the images speckles in the time is a method of monitoring their time variations and consequently the biological activity of the studied object, and, in this context, [12] proposed the Time History of the Speckle Patterns (THSP).

The THSP is a two dimensional image that record a certain line or column of pixels in successive moments and arrange them vertically side by side. The x axis show information about the time evolution of the selected pixels and the y axis is the spatial distribution of the interference patterns [12].

2.2. Co-Occurrence Matrix

The co-occurrence matrix was presented by [13], and expresses the number of the transitions of each THSP pixel with respect to its immediate neighbor. Equation (1) describes mathematically the co-occurrence matrix.

$$M_{CO} = \left[N_{ij}\right] \quad (1)$$

which:

M_{CO} is the co-occurrence matrix, N_{ij} correspond the number of occurrences of an intensity value i, followed by an intensity value j to move through rows or columns of the time history.

Phenomenon that show low biological activities, their time variations of the speckle patterns are slow and present a THSP horizontally in the elongated shape and the co-occurrence matrix is characterized by small changes of the pixels intensity to i and j, as illustrated in the Figure 1(a). However, materials that exhibit high biological activity shows fast intensity variations in the THSP that resemble an ordinary spatial speckle patterns and their co-occurrence matrix has nonzero elements near the main diagonal (**Figure 1**(b)) [14].

2.3. Absolute Values of the Differences (AVD)

One of the methods for analyzing of the speckle patterns is the technique of the absolute values of the differences (AVD), proposed by [15] as an alternative the inertial moment technique.

The AVD method is a statistics moment of first order which it is applied on the co-occurrence matrix and generates a number [11] which allow quantify the biological activity of the studied material. Equation (2) presents mathematically the AVD technique

$$\mathrm{AVD} = \sum_{ij} \left\{ M_{ij} \left| (i-j) \right| \right\} \tag{2}$$

which:

AVD is a dimensionless value, i and j are coordinates of the row and column respectively, and M_{ij} is called of modified co-occurrence matrix and that is presented in Equation (3).

$$M_{ij} = \frac{N_{ij}}{\sum N_{ij}} \tag{3}$$

According [15], the inertial moment showed to be more sensitive than AVD on analyzing processes that involve high biological activities, although when this variation is not so intense, this method is less efficient.

2.4. Fourier Transform

Information of the biospeckle data in the frequency domain has been an alternative to the interpretation of the interference patterns [5], with the possibility of improve the visualization of some phenomena of the studied material and to know their spectral signatures. In this context, the Fourier transform is one of the tools that can be used to spectral analysis of the biospeckle.

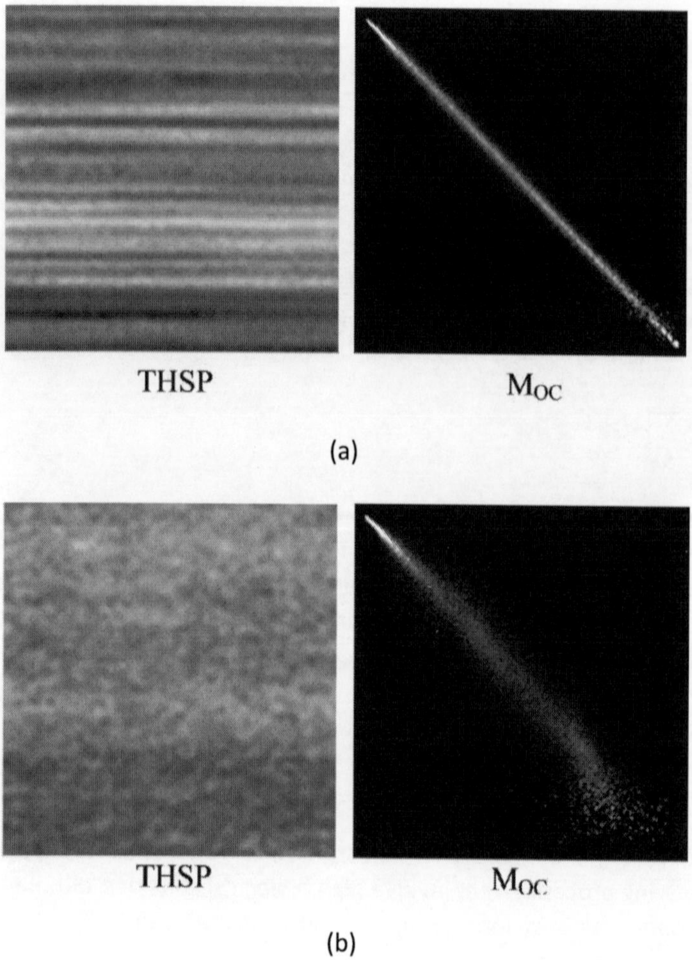

Figure 1. Time history of the speckle patterns and their respective co-occurrence matrix. Materials with low (a) and high (b) biological activity.

Fourier transform can be understood as the mathematical technique that transforms a signal from the time domain to the frequency domain, and it is formed by a set $W_n(t) = e^{int}, n = 0, 1, \cdots$ of orthogonal functions, of period 2π [16]. Equation (4) described mathematically the Fourier transform.

$$f(\omega) = \int_{-\infty}^{\infty} f(t) \cdot e^{-i\omega t} dt \tag{4}$$

which:

$\omega = 2\pi f$

$|f(\omega)|$ = amplitude of each component ω of the signal.

There is also the inverse Fourier transform, which is used to transform the signal from frequency domain to time domain with the reconstruction of the original function. Equation (5) presents the mathematical expression of the inverse Fourier transform.

$$f(t) = \int_{-\infty}^{\infty} f(\omega) \cdot e^{-i\omega t} d\omega \tag{5}$$

The Fourier transform indicates the spectral information of the signal without providing the instant which these components happen, and in situations that to know when the frequencies occur are interesting precludes the use of Fourier transforms, unless if the series is stationary [17]. In this context, the wavelet transform is an alternative that provides the instant the frequency components occur.

2.5. Wavelets Transform

The wavelets are simply waves of duration adjusted with energy concentrated in variables intervals [18], which makes it a great useful method for time series analysis, that exhibit characteristics that can change in the time and in frequency.

The continuous wavelet transform is defined as the convolution of $f(t)$ with a scaled and translated version of ψ [19], called wavelet mother. Equation (6) describes mathematically the continuous wavelet transform.

$$\langle f, \psi_{a,b} \rangle = \int_{-\infty}^{\infty} f(t)\psi\left(\frac{t-b}{a}\right)dt \tag{6}$$

which:

$f(t)$ is the studied signal a scale parameter b translation value

$\psi_{a,b}(t)$ is the mother function of wavelets

$\langle f, \psi_{a,b} \rangle$ is the spectrum wavelets.

The scale is related to the frequency, in which high scales correspond to low frequencies and low scales correspond to high frequencies, whereas the translation is the displacement of the mother function about the studied signal [20].

The return of the signal from frequency domain to time domain, inverse wavelets transform, allows observe the behavior of the signal in specifics frequencies bands and also the reconstruction of the original function $f(t)$. According [19], the inverse wavelet transform can be realized by the sum of real part of wavelet spectrum on all scales (Equation (7)).

$$x_n = \frac{\delta_j \delta_t^{0.50}}{C_\delta \psi_0(0)} \times \sum_{j=0}^{J} \frac{R\{f, \psi_{a,b}\}}{s_j^{0.50}} \tag{7}$$

which:

$s_j^{0.50}$ is a factor that convert the wavelets transform in energy density,

δ_j; $\delta_t^{0.50}$; C_δ; $\psi_0(0)$ are specific constants of the base function used.

One of the major difficulties in wavelet analysis is the identification of the scales set used in the wavelet transform. Orthogonal wavelet, there is a limit and a discrete set of scales, as given by [21], however, for analysis of non-orthogonal wavelet, can use an arbitrary scales set to build a more complete signal [19].

In this context, [19] suggested Equations (8) and (9) to calculate the scales interval to be used in the wavelet transform, in which s_j is the lowest and J is the highest scale.

$$s_j = s_0 \cdot 2^{j \cdot \delta j} \; j = 0,1,2,\cdots,J \qquad (8)$$

$$J = \delta j^{-1} \cdot \mathrm{LOG}\left(\frac{N \cdot \delta t}{S_0}\right) \qquad (9)$$

The s_0 should be chosen so that the Fourier period is $2 \cdot \delta t$, and to the Morlet wavelet the largest value that can adjust the scale is δj of 0.5. For other wavelet functions can be used a larger value.

2.6. Sampling Theorem

The sampling theorem describes the relationship between sampling frequency of a signal and the frequency maximum of the reconstructed signal. Below is transcript the sampling theorem as presented by [22].

"Theorem 1: If a function $f(t)$ contains no frequencies higher than W cps, it is completely determined by giving its ordinates at a series of points spaced 1/2 seconds W apart".

According to the theorem, the number of samples per unit time of a signal is called rate or frequency sampling (W), and half the sampling frequency corresponds to the frequency maximum of the signal which can be reproduced in full without aliasing error.

The sampling theorem is used in this work to define the highest frequently during the decomposition of signals.

3. MATERIALS AND METHODS

It was conducted a comparison between Fourier and wavelet transforms using the time history of speckle patterns (THSP) relative to a paint drying process and presented by [23].

The database was formed by 8 THSP's collected each 20 minutes during the paint drying using the back-scattering experimental setup. Each time history was made by a set of 128 images, resolution of 512 by 640 pixels, whose time acquisition between images was of 0.08 seconds (sampling frequency of 12.5 Hz).

The lines of the THSPs were concatenated creating a new signal that was decomposed into frequency spectra using Fourier and wavelet transforms with application posterior of the inverse transform. Some frequency bands were eliminated before the reconstruction of the signal in order to analyze the results of the speckle signal using a numerical method to measure the speckle activity. The selective filtering was conducted as well in order to create some frequency markers linked to the physical phenomena under monitoring.

According the sampling theorem the highest frequency that can be seen in the reconstruction process is 6.25 Hz, and using Equations (8) and (9) were calculated the number of frequency bands used in the transform. In addition, in the continuous wavelet transform was used mother function of Morlet, a damped complex exponential with a set of oscillation parameter that preserves an approximate relationship between the scale of the wavelet analysis and the frequency in a Fourier analysis, as described by [24].

The signal resulting of the inverse transform was converted to THSP format again and numerically analyzed using the technique of the absolute values of the differences (AVD) [15], and their values compared to the gravimetrical measurement.

Figure 2 illustrated all the methodology used.

4. RESULTS AND DISCUSSION

Figure 3 presents the absolute value of the differences for the THSP's of the paint drying process with decomposition and reconstruction of some frequency bands using Fourier transform.

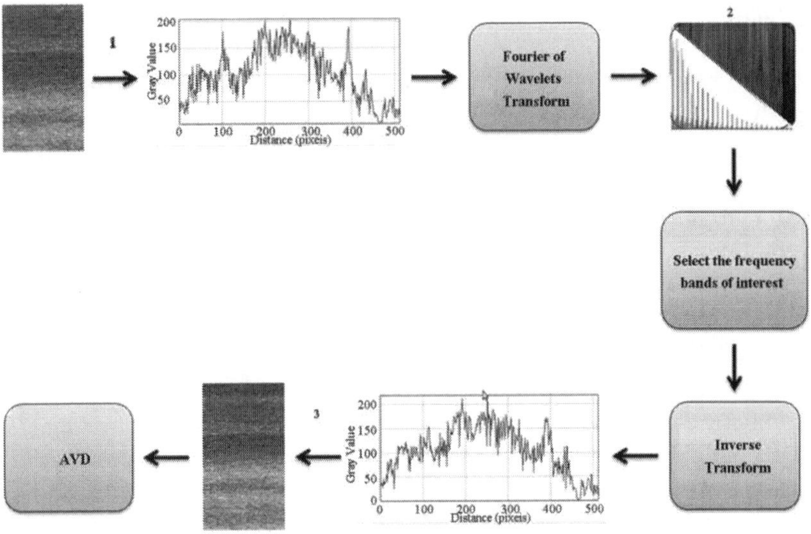

Figure 2. Methodology used to the data analysis, in which 1 represents the concatenation, 2 is the Fourier or wavelets spectrum and 3 corresponds to the inverse process of the concatenation.

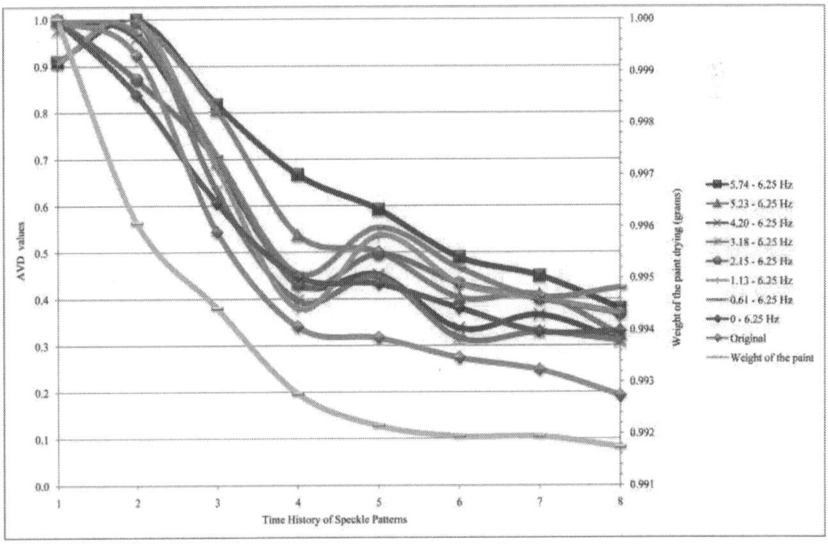

Figure 3. AVD values of the THSP's of the paint drying using Fourier transform for spectral analysis.

The data reconstructed in the frequency band from 0 to 6.25 Hz (**Figure 3**), here called of total reconstruction, correspond to highest reconstruction possible in accordance with the sampling theorem, and the original data are the AVD values of the time history of the paint drying without filtering.

The dynamic speckle signals reconstructed in frequencies bands 5.74 - 6.25 and 5.23 - 6.25 Hz presented gradual reduction of AVD values along of the paint drying process, closer to the behavior of the original data, total reconstruction and to the weight of the paint drying. However, the addition of low frequencies components in the reconstruction process resulted in the oscillation of the AVD values in the fifth time history, as observed in Figure 3. We attribute those oscillations to the influence of the atmospheric conditions that occur in experiment of paint drying as the realized by [23], which did not interfered in the first moments since the paint volatility was higher and thus undermined the presence of the modulation of the signal in low frequencies linked to the atmospheric conditions such as temperature and humidity, and **Figure 4** illustrates these information's.

Figure 4 presents that the signals reconstructed with components of low frequencies (associated the temperature and humidity variables) are mixed with the data reconstructed using components of high frequencies (linked the volatility) in the first moments, and which along the paint drying process, the paint volatility stabilize and make possible to observe the high oscillations of the data reconstructed using low frequencies component.

In order to clarify the spectral information found in the fifth THSP, a mathematical model was adjusted in the AVD values of the original data, using the least squares method, to describe the process of paint drying. Equation (10) presents the mathematical model adjusted and Figure 5illustrates the regression curve.

$$y = -0.003624x^4 + 0.06257x^3 - 0.340004x^2 + 0.450417x + 0.889444 \quad (10)$$

which:

x is the TSHP numbery is the normalized AVD value.

The mathematical model adjusted showed a correlation index of 0.98 and mean square error of 0.0048 with respect to the original.

The first derivative of the adjusted model,

$$y' = -0.014496x^3 + 0.188869x^2 - 0.680007x + 0.450417$$

when equaled to zero, showed that the region the fifth THSP is a minimum local, and this means that the AVD values reaches a minimum value and then initiates an oscillations related to the variations of the temperature and humidity. These results are similar to the behavior of the paint weight at time presented by [23] (**Figure 5**), in which is possible to observe a stabilization of the weight values after the fifth data acquisition, and leads us to assume that the drying time of this paint is one hour and twenty minutes, approximately. In this context, the dynamic speckle analyzed by Fourier transform allowed to observe this transition occurred in paint drying structure.

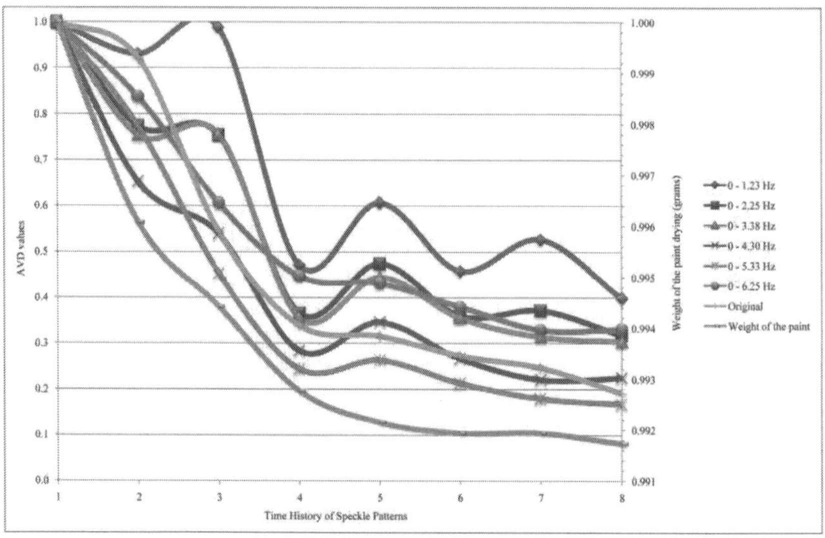

Figure 4. AVD values of the time history of the paint drying reconstructed with some frequencies band using Fourier transform.

In addition, the distance between the original signals and reconstructed signals were evaluated and the results are illustrated in **Figure 6**.

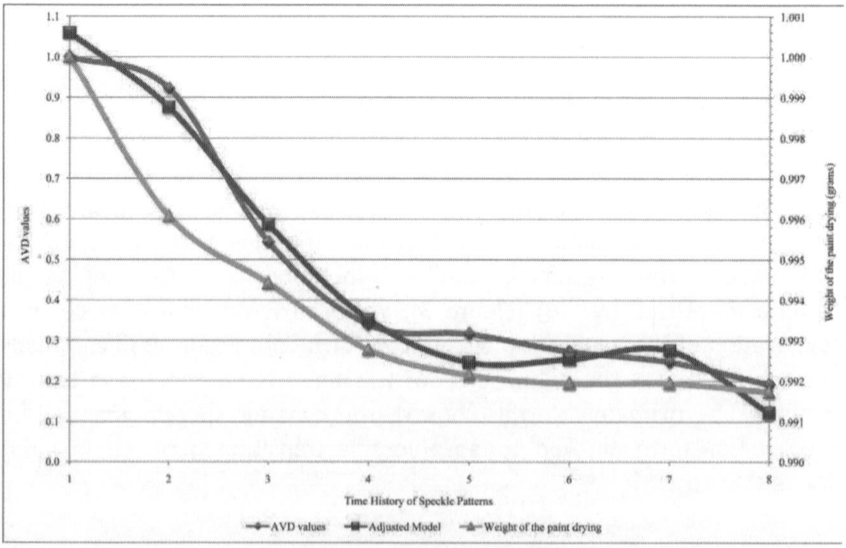

Figure 5. Regression adjusted to describe the paint drying process.

The largest distances were observed in the signals reconstructed using few spectral components and these distances between original signals and reconstructed signals were reducing to use a larger number of components in the inverse Fourier transform, as waited.

In addition, the correlation index between the original signal and the signals reconstructed in the frequencies bands of 5.74 - 6.25, 5.23 - 6.25 and 4.20 - 6.25 Hz were of 0.86, 0.94 and 0.98, respectively. The high correlation of 0.86 using just a small portion of the frequency band can be explained by compact support of the Fourier basis functions in the frequency domain, which allows using in data compression with minimum loss of information, as discussed by [16].

The speckle signals is not stationary [25], unviable to use the Fourier transform. Thus, frequency analysis using the Fourier transform is restricted to situations in which are interesting to know the spectral information of the data.

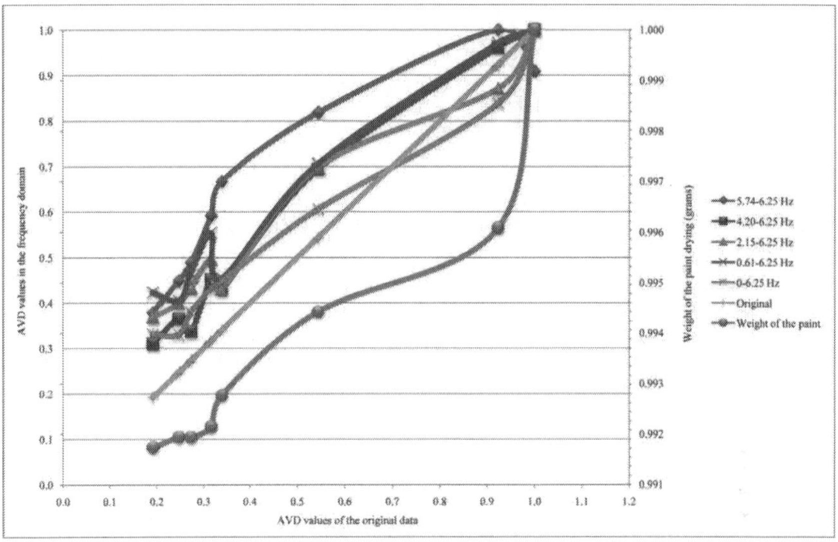

(a)

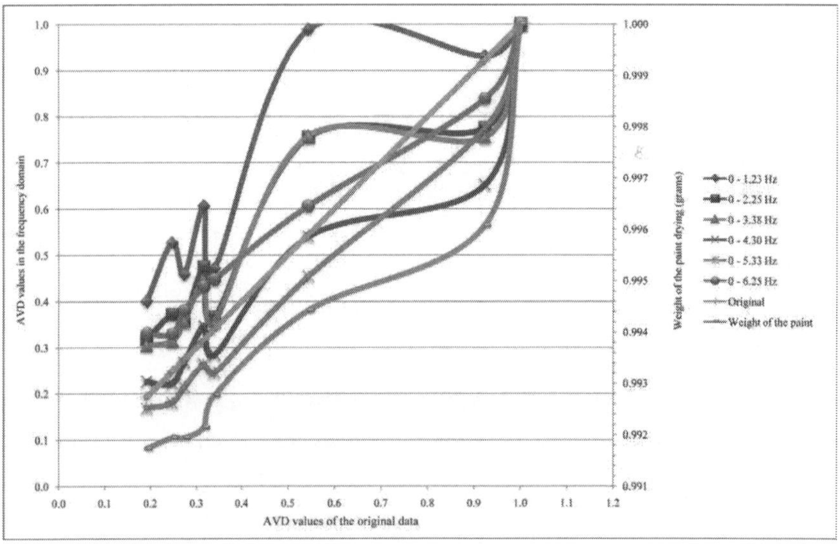

(b)

Figure 6. Signals reconstructed in specific frequency bands by inverse Fourier transform and the original signal. (a) Addition of components of low frequencies in the signals reconstruction and (b) increase high frequencies components in the inverse transform.

Otherwise, the spectral analysis using the wavelet transform presented different behavior in the high frequencies and the signal reconstructed using more components of low frequencies made the results closer to the original signal, as illustrated in **Figure 7**.

The total reconstruction of the biospeckle signal, in the frequency band from 0 to 6.25 Hz, showed behavior similar to the original data and to the weight of paint drying, with the gradual reduction of the AVD values in the time.

Furthermore, removing the low frequency components in the reconstruction process resulted in oscillations in the AVD values, in special, in the signals reconstructed within the frequency range from 4.20 to 6.25 Hz, which the AVD values decreasing until the fourth time history and subsequently increasing.

In this context, the high AVD values of the last time history in the frequency bands 4.20 - 6.25 Hz are attributed the random oscillations and noise presents in the biospeckle signal, without significant information's about the paint volatility, since that the energy of the time history in the high frequencies showed was reducing along of the paint drying and presented low values after of the fourth THSP, as illustrated graphically in **Figure 8**.

The energy of the THSP's (**Figure 8**) is represented in pseudo-colors, the ordinate axis correspond the scales and in the abscissas axis is the time. The light pseudocolors indicate high energy while dark shades are associated the low energy, and the scales are inversely proportional to the frequencies, which the low scales are attached with high frequencies and high scales with the low frequencies.

It is possible to observe in **Figure 8** that the pseudocolors in the high frequencies (low scales) are darkening in the time, which means reduce of the energy in high frequency bands along of the paint drying and which does not justify the high AVD values in the last time histories.

The reconstruction of the signals using components of low frequencies was also analyzed, and**Figure 9** shows the results.

In the first moments we see that the signals reconstructed using components of low frequency and of high frequency mixed, instant in which the paint volatility was intense. Over time, the phenomena linked to the high frequencies stabilized, allow to observe the oscillations of the signals reconstructed with components of low frequency, as the humidity and the temperature.

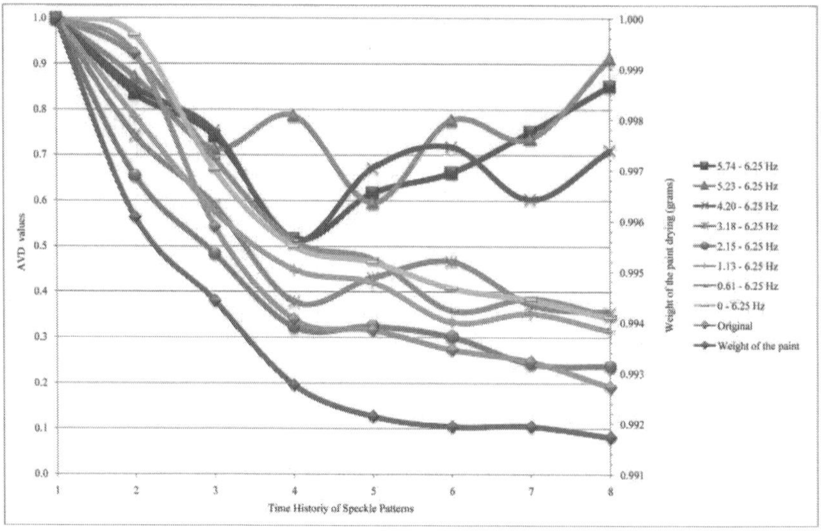

Figure 7. AVD values of the time history of speckle patterns of paint drying using wavelet transform for spectral analysis.

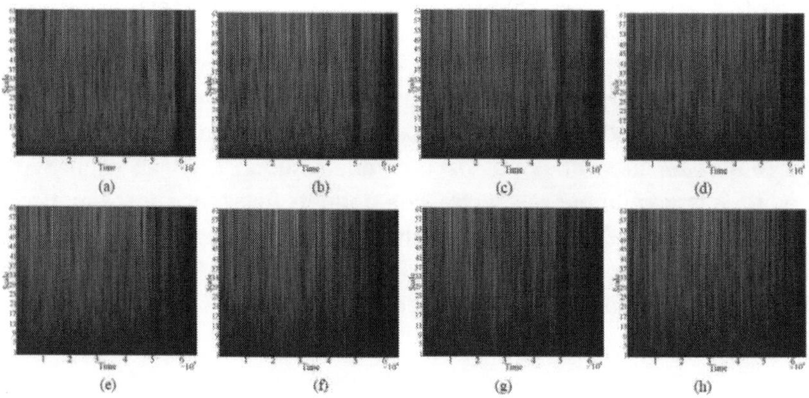

Figure 8. Energy of the 8 THSP's for different frequencies.

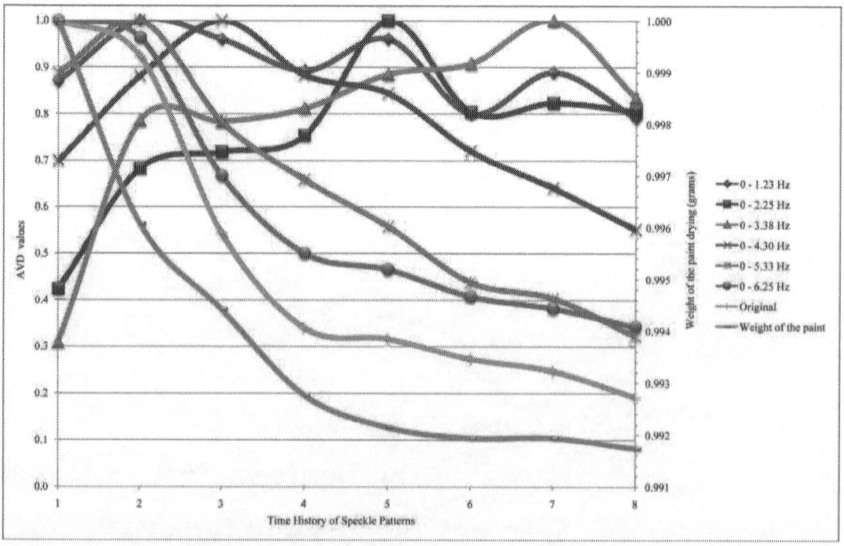

Figure 9. Absolute value of the difference of the THSP's of the paint drying using wavelet transform for spectral analysis.

The reconstruction of the signals showed correlation index higher than 0.90 with the original signal when used a wide frequency band, the opposite showed when we used the Fourier transform.**Figure 10** presents the signals reconstructed using wavelet transform against the original signal, and allow observation the distance between the curves when we added components of high and low frequencies in the reconstruction of the signals.

Signals reconstructed in frequency bands of 5.23 - 6.25, 4.20 - 6.25 and 3.18 - 6.25 Hz presented correlation index of 0.30, 0.69 and 0.92, respectively. The Morlet function does not have compact support, which explains the need for a large number of spectral components for great approximation of the original signal. Furthermore, the addition of components of low frequencies (**Figure 10**(a)) and of high frequencies (**Figure 10**(b)) in the signal reconstruction made the reconstructed data closer to the original signal.

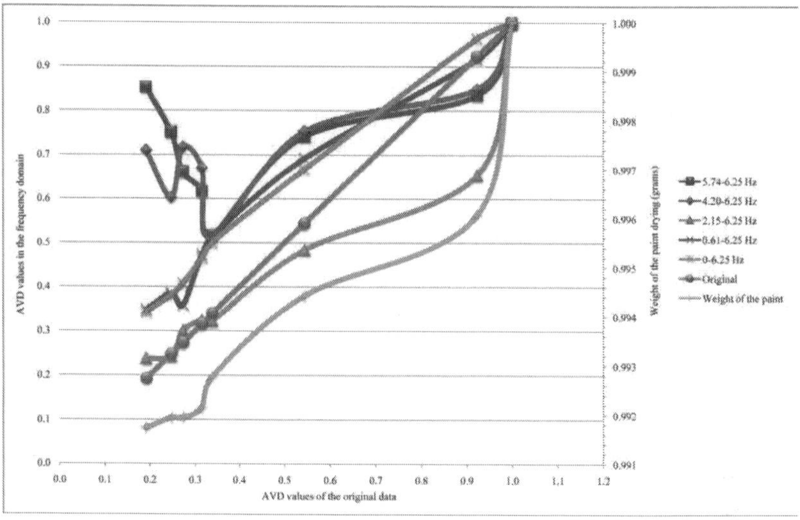

(a)

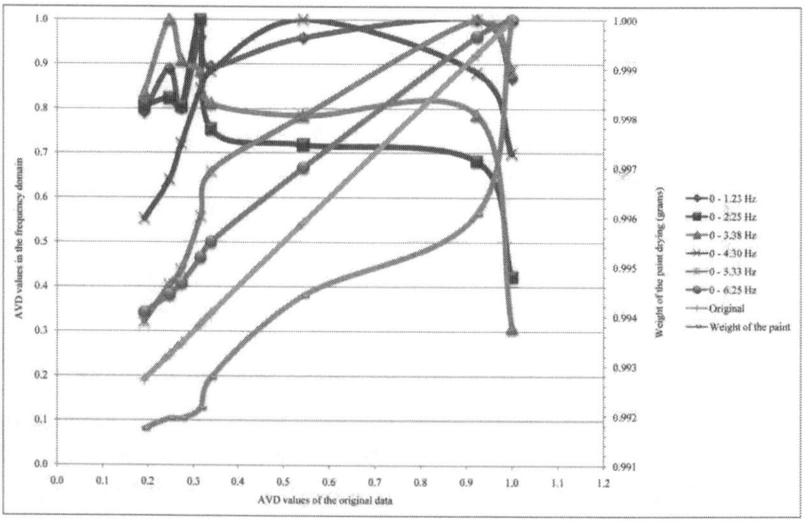

(b)

Figure 10. Signals reconstructed using wavelet transform and the original signal. (a) Addition of components of low frequencies in the inverse wavelet transform and (b) increase high frequencies components in the signal reconstruction.

In this context, the wavelet transform details spectral information's in time, which does not occur in the Fourier transform, and that in the analysis of the biospeckle signals provides further information of the studied process and facilitates to understand the signals that create this complex phenomenon of optical interference, being the adequate tool for studies in the frequency domain.

5. CONCLUSION

The Fourier transform allowed data analysis with compact support, while the wavelets provided definition of frequency markers and information is not presented in the Fourier analysis about the participants of the dynamic speckle signal and being the tool that most adequate for frequency analysis.

ACKNOWLEDGEMENTS

This work was partially financed by CNPq, Fapemig, Capes, Finep in Brazil, and partly supported by the Scottish Government Rural and Environment Science and Analytical Services division.

REFERENCES

1. Y. Zhao, J. Wang, X. Wu, F. W. Williams and R. J. Schmidt, "Point-Wise and Whole-Field Laser Speckle Intensity Fluctuation Measurements Applied to Botanical Specimens," Optics and Lasers in Engineering, Vol. 28, No. 6, 1997, pp. 443-456.

2. H. J. Rabal and R. A. Braga, "Dynamic Laser Speckle and Applications," CRC Press, New York, 2008.

3. A. P. Rathnayake, H. Kadono, S. Toyooka and M. Miwa, "A Novel Optical Interference Technique to Measure Minute Root Elongations of Japanese Red Pine (Pinusdensiflora Seibold & Zucc.) Seedlings Infected

with Ectomycorrhizal Fungi," Environmental and Experimental Botany, Vol. 64, No. 3, 2008, pp. 314-321.

4. R. A. Braga, L. Dupuy, M. Pasqual and R. R. Cardoso, "Live Biospeckle Laser Imaging of Root Tissue," European Biophysics, Vol. 38, No. 5, 2009, pp. 679-686.

5. R. R. Cardoso, A. G. Costa, C. M. B. Nobre and R. A. Braga, "Frequency Signature of Water Activity by Biospeckle Laser," Optics Communications, Vol. 284, No. 8, 2011, pp. 2131-2136.

6. A. Zdunek and W. B. Herppich, "Relation of Biospeckle Activity with Chlorophyll Content in Apples," Postharvest Biology Technology, Vol. 64, No. 1, 2012, pp. 58-63.

7. I. Passoni, A. Dai Pra, H. J. Rabal, M. Trivi and R. Arizaga, "Dynamic Speckle Processing Using Wavelets Based Entropy," Optics Communications, Vol. 246, No. 1-3, 2005, pp. 219-228.

8. R. M. Costa, T. Sáfadi, G. F. Rabelo and R. A. Braga, "Té- cnicas Estatísticas Aplicadas em Imagens do Speckle Dinâmico," Revista Brasileira de Biometria, Vol. 28, No. 2, 2010, pp. 27-39.

9. G. F. Rabelo, A. M. Enes, R. A. Braga and I. M. Dal Fabbro, "Frequency Response of Biospeckle Laser Images of Bean Seeds Contaminated by Fungi," Biosystems Engineering, Vol. 110, No. 3, 2011, pp. 297-301.

10. G. H. Sendra, R. Arizaga, H. Rabal and M. Trivi, "Decomposition of Biospeckle Images in Temporary Spectral Bands," Optics Letters, Vol. 30, No. 13, 2005, pp. 1641- 1643.

11. M. D. Z. Ansari and A. K. Nirala, "Assessment of the BioActivity Using the Methods of Inertia Moment and Absolute Value of the Differences," Optiks—International Journal for Light and Electron Optics, Vol. 124, No. 15, 2013, pp. 2180-2186.

12. G. Oulamara, J. Tribillon and J. Duvernoy, "Biological Activity Measurements on Botanical Specimen Surfaces Using a Temporal Decorrelation Effect of Laser Speckle," Journal of Moderns Optics, Vol. 36, No. 2, 1989, pp. 136-179.

13. R. Arizaga, M. Trivi and H. Rabal, “Speckle Time Evolution Characterization by the Co-Occurrence Matrix Analysis,” Optics and Laser Technology, Vol. 31, No. 2, 1999, pp. 163-169.

14. R. A. Braga, I. M. Dal Fabbro, F. M. Borém, G. F. Rabelo, R. Arizaga, H. Rabal and M. Trivi, “Assessment of Seed Viability by Laser Speckle Techniques,” Biosystems Engineering, Vol. 86, No. 3, 2003, pp. 287-294.

15. R. A. Braga, C. M. B. Nobre, A. G. Costa, T. Sáfadi and F. M. Costa, “Evaluation of Activity through Dynamic Laser Speckle Using the Absolute Value of the Differences,” Optics Communications, Vol. 284, No. 2, 2011, pp. 646- 650.

16. P. A. Morettin, “Ondas e Ondeletas,” EDUSP, São Paulo, 1999.

17. M. Sifuzzaman, M. R. Islam and M. Z. Ali, “Application of Wavelet Transform and Its Advantages Compared to Fourier Transform,” Journal of Physical Sciences, Vol. 13, No. 1, 2009, pp. 121-134.

18. A. Graps, “An Introduction to Wavelets,” Computation Science and Engineering, IEEE, Vol. 2, No. 2, 1995, pp. 50-61.

19. C. Torrence and G. P. Compo, “A Practical Guide to Wavelet Analysis,” Bulletin of the American Meteorological Society, Vol. 79, No. 1, 1998, pp. 61-78.

20. H. R. Karimi, W. Pawlus and K. G. Robbersmyr, “Signal Reconstruction, Modeling and Simulation of a Vehicle Full-Scale Crash Test Based Morlet Wavelets,” Neurocomputing, Vol. 93, No. 15, 2012, pp. 88-99.

21. M. Farge, “Wavelet Transforms and Their Applications to Turbulence,” Annual Review of Fluid Mechanics, Vol. 24, No. 1, 1992, pp. 395-457.

22. C. E. Shannon, “Communication in the Presence of Noise,” Proceedings of the IRE, Vol. 37, No. 1, 1949, pp. 10-21.

23. M. M. Silva, J. R. A. Nozela, M. J. Chaves, R. A. Braga and H. Rabal, “Optical Mouse Acting as Biospeckle Sensor,” Optics Communications, Vol. 284, No. 7, 2011, pp. 1798-1802.

24. L. Polansky, G. Wittemyer, P. C. Cross, C. J. Tambling and W. M. Getz, "From Moonlight to Movement and Synchronized Randomness: Fourier and Wavelet Analyses of Animal Location Time Series Data," Ecology, Vol. 91, No. 5, 2010, pp. 1506-1518.

25. H. Sendra, S. Murialdo and L. Passoni, "Dynamic Laser Speckle to Detect Motile Bacterial Response of Pseudomonas Aeruginosa," Journal of Physics: Conference Series, Vol. 90, No. 1, 2007, pp. 1-6.

Chapter 4

Computerized Heart Sounds Analysis

S.M. Debbal[1]

[1] Genie -Biomedical Laboratory (GBM), Department of electronicFaculty of science engineering university Aboubekr Belkaid , Algeria

1. INTRODUCTION

Heartbeat sound analysis by auscultation is still insufficient to diagnose some heart diseases. It does not enable the analyst to obtain both qualitative and quantitative characteristics of the phonocardiogram signals [1],[2]. Abnormal heartbeat sounds may contain, in addition to the first and second sounds, S1 and S2, murmurs and aberrations caused by different pathological conditions of the cardiovascular system [2]. Moreover, in studying the physical characteristics of heart sounds and human hearing, it is seen that the human ear is poorly suited for cardiac auscultation [3]. Therefore, clinic capabilities to diagnose heart sounds are limited.

The sound emitted by a human heart during a single cardiac cycle consist of two dominant events, known as the first heart sound S1 and the second heart sound S2 (Figure1), S1 relates to the closing of the mitral and tricuspid valves whilst S2 is generated by the halting of the aortic and pulmonary valves leaflets [1]. S1 corresponds in timing to the QRS complex in ECG (Electrocardiogram) and S2 follows the systolic pause in the normal cardiac cycle. Heartbeat sound analysis by auscultation only is still insufficient to diagnose some heart diseases. It does not enable the analyst to obtain both qualitative and quantitative characteristics of S1 and S2 of the phonocardiogram [1-2]. Moreover, in studying the physical characteristics of heart sounds and human hearing, it is seen that the human ear is poorly suited for cardiac auscultation [3]. Therefore, clinic capabilities to diagnose heart sounds are limited. In this paper we are interested in the study of the physical characteristics of *the second heart sound S2 which consist of two major components in the spectrum of the signal*. One of these components A2 is due to the closure of the aortic valve and the other P2 is due to the closure of the pulmonary valve.

The aortic component is loudest than the pulmonic component. It is discernible at all the auscultation sites.It is best heard at the right base, with the diaphragm of the chest piece firmly pressed, whereas the pulmonic on a may only be audible at the left base, with the diaphragm of the chest piece firmly pressed.

The aortic component has higher frequency contents and generally precedes the pulmonary component because in normal heart activity the aortic valve closes before the pulmonary valve. The difference of time occurrence between these valves activities is known in medical community [4 - 6] as split. However in many diseases [1] this order of time occurrence may be reverse and its delay varies.

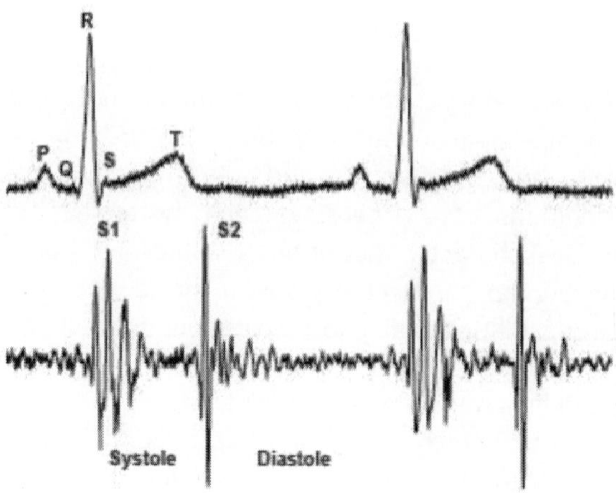

Figure 1. Correlation between the phonocardiogram signal (PCG) and the electrocardiogram signal (ECG)

The characteristics of the PCG signal and other features such as heart sounds S1 and S2 location; the number of components for each sound; their frequency content; their time interval; all can be measured more accurately by digital signal processing techniques.

The FFT (Fast Fourier Transform) can provide a basic understanding of the frequency contents of the heart sounds. However, FFT analysis remains of limited values if the stationary assumption of the signal is violated. Since heart sounds exhibit marked changes with time and frequency, they are therefore classified as non - stationary signals. To understand the exact feature of such signals, it is thus important, to study their time – frequency characteristics.

In this paper the wavelet transform is used to analyse both the normal and abnormal heart sound in both time and frequency domains. This technique has been shown to have a very good time resolution for high-frequency components. In fact the time resolution increases as the frequency increases and the frequency resolution increases as the frequency decreases [4],[5].

Furthermore, the wavelet transform has demonstrated the ability to analyse the heart sound more accurately than other techniques STFT or Wigner distribution [6] in some pathological cases.

In fact the spectrogram (STFT) cannot track very sensitive sudden changes in the time direction. To deal with these time changes properly it is necessary to keep the length of the time window as short as possible. This however, will reduce the frequency resolution in the time-frequency plane. Hence, there is a trade-off between time and frequency resolutions [6].

However the Wigner distribution (WD) and the corresponding WVD (Wigner Ville Distribution) have shown good performances in the analysis of non-stationary signals. This comes from the ability of the WD to separate signals in both time and frequency directions. One advantage of the WD over the STFT is that it does not suffer from the time-frequency trade-off problem. On the other hand, the WD has a disadvantage since it shows cross-terms in its response. These cross-terms are due to the nonlinear behaviour of the WD, and bear no physical meaning. One way to remove these cross-terms is by smoothing the time-frequency plane, but this will be at the expense of decreased resolution in both time and frequency [7].

The WD was applied to heart sound signal it shows no succes in displaying or separating the signal components in both the time and frequency direction [6], although it provides high time-and frequency- resolution in simple monocomponent signal analysis[8].

To overcome these difficulties with the STFT and the WD an alternative way to analyse the non-stationary signals is the wavelet transform (WT). It expand the signal some basis functions. The basis functions can be constructed by dilation, contractions and shifts of a unique function called the "wavelet prototype" or "wavelet mother". The WT act as "mathematical microscope" in which we can observe different parts of the signal by just adjusting the focus.

The wavelet Transform is a technique in the domain of time-frequency distributions. The main idea of this method is the representation of an arbitrary signal as a superposition of basic signals, "atoms", located in time and frequency. These atoms may be derived by means of a special operation on a single parent atom. Parent atoms and derivation operation are usually chosen such as to enable the construction of an orthonormal system [9].

The study of the decomposition of the signal in " atoms " was first carried out by Gabor however, it was quickly abandoned be cause of :

1) the no simultaneous Representation in time and frequency

2) grid made up of rectangular cells is not a flexible device

3) the mathematical theory of the phenomenon is badly structured.

The representation time-scale of WT based on a dyadic paving appears more flexible. It a mathematical structure governed by a formula of exact inversion [10] making possible the existence of orthonormal basis. This makes the wavelet to be a simultaneous function of time and frequency.

In this paper the continuous wavelet transform (CWT) is applied to analyse pathological PCG signals. The CWT is more appropriate than the discrete wavelet transform (DWT), since we are interested in the analysis of non-stationary signals and not signal coding where DWT is found to be more useful

2. THEORETICAL BACKGROUND

2.1. Fourier Transform (Ft)

In 1882, Joseph Fourier discovered that any periodic function could be represented as an infinite sum of periodic complex exponential functions [10]. The inclusive property of only periodic functions was later extended to any discrete time function. The Fourier transform (FT) [as regular Fourier Transform] converts a signal expressed in the time domain to a signal expressed in the frequency domain. The FT is widely used and usually implemented in the form of FFT algorithm (fast Fourier transform). The mathematical definition of the FT is given below.

$$X(f) = \int x(t)e^{-j2\Pi ft}dt \tag{1}$$

Where t and f are respectively the time and frequency parameters. The time domain signal x(t)) is multiplied by a complex exponential at a frequency ffand integrate over all time. In other words, any discrete time signal may be represented by a sum of sines and cosines, which are shifted and are multiplied by a coefficient that changes their amplitude. X(f)are the Fourier coefficients which are large when a signal contains a frequency component around the frequency f The peaks in a plot of the FT of a signal correspond to dominant frequency components of the signal. Fourier analysis is simply not effective when used on non stationary signals because it does not provide frequency content information localized in time. Most real world signals exhibit non stationary characteristics (such as heart sound signals). Thus, Fourier analysis is not adequate.

2.2. Short-Time Fourier Transform (Stft)

The STFT is obtained by calculating the Fourier transform of a sliding windowed version of the time signal s(t). The location of the sliding window adds a time dimension and one gets a time-varying frequency analysis.

The mathematical representation of STFT is :

$$S(t,f)=\int_{-\infty}^{+\infty} s(\tau)w(\tau-t)e^{-j2\pi\ ft}d\tau \tag{2}$$

Where w(τ−t) it is the sliding window applied to the signals(t),f is the frequency and t is the time.The length of the window is chosen so that to maintain signal stationary in order to calculate the Fourier transform. To reduce the effect of leakage (the effect of having finite duration), each sub-record is then multiplied by an appropriate window and then the Fourier transform is applied to each sub-record. As long as each sub-record does not contain rapid changes the spectrogram will give an excellent idea of how the spectral composition of the signal has changed during the whole time record. However, there exist many physical signals whose spectral content is so rapidly changing that finding an appropriate short-time window is problematic, since there may not be any time interval for which the signal is stationary. To deal with these time changes properly it is necessary to keep the length of the time window as short as possible. This, however, will reduce the frequency resolution in the time-frequency plane. Hence, there is a trade-off between time and frequency resolutions.

2.3. Wgner Distribution Function (Wd)

In contrast to the STFT, which is resolution limited either in time or in frequency (dictated by the window function), and suffers from smearing and side lobe leakage, the WD offers excellent resolution in both the frequency and time domains. The WD of two signals, $x(t)$, $y(t)$, is defined via,

$$W(t,\omega)=\int_{-\infty}^{+\infty} x(t+\frac{\tau}{2})y^*(t-\frac{\tau}{2})e^{-j\omega\tau}d\tau \tag{3}$$

The auto WD is obtained when $x(t)$ = $y(t)$=s(t), it is a bilinear function of the signal s(t). the WD, and can also be expressed as

$$S(t,\omega)=\int_{-\infty}^{+\infty} s(t+\frac{\tau}{2})s^*(t-\frac{\tau}{2})e^{-j\omega\tau}d\tau \tag{4}$$

Where t and ω are respectively, the time and frequency variables, and * denotes the complex conjugate.

The WD had shown good applications in the analysis of non-stationary signal [7]. This comes from the ability of this method to separate signals in both time and frequency directions. The WD has a disadvantage that it is limited by the appearance of cross-terms. These cross-terms are due to the nonlinear behaviour of the WD, and bear no physical meaning. One way to remove these cross-terms is by smoothing the time-frequency plane [7], but this will be at the expense of decreased resolution in both time and frequency.

2.4. Continuous Wavelet Transform (Cwt)

The continuous wavelet transform was developed as a method to obtain simultaneous, high resolution time and frequency information about a signal. The term 'wavelet' was first mentioned in 1909 in a thesis by Alfred Haar [M.Misiti.Y.Misiti.G.Oppenheim.J.-M.Poggi.Wavelet Toolbox:For use with MATLAB.The Math Works Inc.1996], although the progress in the field of wavelet has been relatively slow until the 1980's when scientist and engineers from different fields realized they were working on the same concept and began collaborating.

The CWT rather than the STFT uses a variable sized window region.Because the wavelet may be dilated or compressed, different features of the signal are extracted. While a narrow wavelet extracts high frequency components, a stretched wavelet picks up on the lower frequency components of the signal.

The CWT is computed by correlating the signal s(t) with families of time-frequency atoms g(t), it produce a set of coefficients C(a,b) given by :

$$C(a,b) = \frac{1}{\sqrt{a}} \int_{-\infty}^{+\infty} s(t) g^* \left(\frac{t-b}{a}\right) dt \tag{5}$$

$$= \sqrt{a} \int_{-\infty}^{+\infty} G(aw)S(w)e^{jbw}dw \qquad (6)$$

Where

- b is the time location
- a is called scale factor and it is inversely proportional to the frequency (a > 0)
- *dénotes a complexe conjugate.
- g(t) is the analysing wavelet.
- S(w) and G(w) are, respectively, the Fourier transforms of s(t) and g(t).

The analyzing wavelet function g(t) should satisfy some properties. The most important ones are continuity, integrability, square integrability, progressivity and it has no d.c component. Moreover, the wavelet g(t) has to be concentrated in both time and frequency as much as possible. It is well known that the smallest time-bandwidth product is achieved by the Gaussian function [3]. Hence the most suitable analyzing wavelet for time-frequency analysis is the complex exponential modulated Gaussian function of the form :

$$g(t) = \exp(-\frac{t^2}{2} + j\, w_o\, t) \qquad (7)$$

If we choose the analyzing wavelet that has the following Fourier transform (FT):

$$G(w) = A\exp[-(w-w_o)^2/2] + \varepsilon \qquad (8)$$

When ε is a small correction term, theoretically necessary to satisfy the admissibility conditions of wavelets; wo is chosen large enough so that the correction term is negligible and can be ignored.

This is known as the Gabor wavelet. It was shown [5] that wo = 5.33, which is enough to make the correction term negligible and gives an optimal time-bandwidth product.

In a continuous wavelet transform, the wavelet corresponding to the scale and the time location b is given by

$$g_{a,b}(t) = 1/\sqrt{|a|}\ g\left(\frac{t-b}{a}\right) \tag{9}$$

Where g(t) is the wavelet "prototype" or mother which can be thought of as a band pass function. The factor /a/−1/2/is used to ensure energy preservation [5].

2.5. Discrete Wavelet Transform (Dwt)

Wavelet series (WS) coefficients are sampled CWT coefficients. Time remains continuous but time-scale parameters (b,a) are sampled on a so-called "dyadic" grid in the time-scale plane (b,a) [11]. A common definition is :

$$C_{jk} = \mathrm{CWT}\{s(t);\ a = 2^j,\ b = k\,2^j\}\quad \text{with } j,k \in Z \tag{10}$$

The wavelets are in this case :

$$\Psi_{jk}(t) = 2^{-j/2}\ \Psi(2^{-j}t - k) \tag{11}$$

The discrete wavelet transform (DWT) has been recognized as a natural wavelet transform for discrete-time signals. Both the time and time-scale parameters are discrete.

The discretization process partially depends upon the algorithm chosen to perform the transformation.

The Cj,k could be well approximated by digital filter banks. By using Mallat's [12]remarkable fast pyramid algorithms which involve use of low-pass and high-pass filters. The Mallat algorithm is in fact a classical scheme known in the signal processing community as two-channel subband coder. The wavelet analysis permits to decompose the phonocardiogram signal in filter banks. The signal to be analyzed is decomposed in approximation an in detail while using two filter h and g. The h filter is a low - pass filter with a band pass [0,Π/2], it generates the approximation signal "A".

The g filter is a high – pass filter of band pass [0, Π], it generates the detail signal "D". The filtered signals are under-sampled (decimation) according to the rule of Nyquist. This leads to a considerable reduction in computing time. Every approximation is decomposed all over again in approximation and detail. In this case the signal is decomposed into several frequency bands instead of two bands. The number of bands depends on decomposition level.

Figure2 shows the discrete wavelet decomposition on three levels, as well as the note associated filter bank for each level. We note that the width of the filter banks band-pass is veritable according to the decomposition level. The wavelet analysis permits to decompose the phonocardiogram signal in filter banks. The signal to be analyzed is decomposed in approximation an in detail while using two filter h and g. The h filter is a low - pass filter with a band pass [0,Π/2], it generates the approximation signal "A".

2.6. Wavelet Packet Transform (Wpt)

Wavelet packet analysis is an extension of the discrete wavelet transform (DWT) and it turns out that the DWT is only one of the much possible decomposition that could be performed on the signal. Instead of just decomposing the low frequency component, it is therefore possible to subdivide the whole time-frequency plane into different time-frequency pieces as can be seen from. The advantage of wavelet packet analysis is that it is possible to combine the different levels of decomposition in order to achieve the optimum time-frequency representation of the original [13].

To obtain a uniform filter bank we will use the wavelet packet analysis which is a generalization of the discrete wavelet transform analysis. The decomposition is also made of the details signal "D". Figure3shows the wavelet packet decomposition on three levels as well as the filter banks associated for each uniform level.

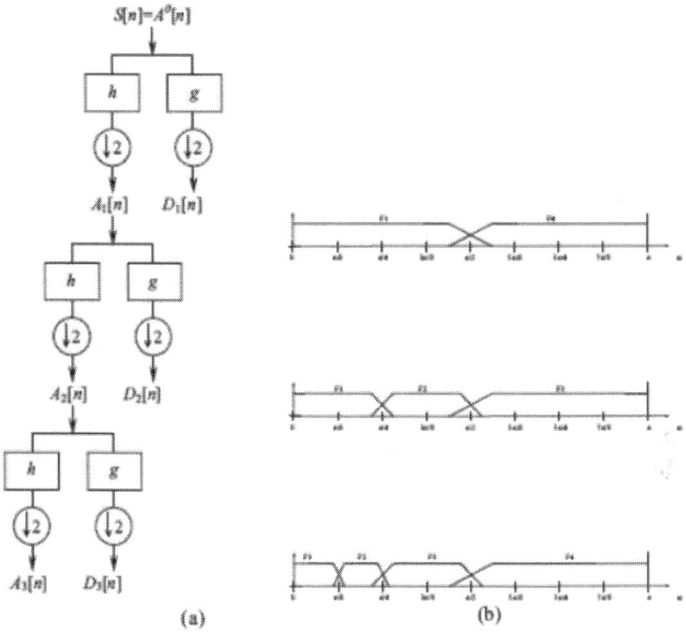

Figure 2. a) Exemple of the discrete wavelet decompostion for three levels; b) Filter banks of the decomposition of each level

3. RESULTS AND DISCUSSION

3.1. Fast Fourier Transform Application

An FFT algorithm is first applied to the PCG signal given in Figure1. The frequency spectrum illustrated in Figure4a shows that the normal PCG signal has a frequency content varying from around 40Hz up to 200Hz. The FFT can be applied to the first part of this signal to analyse the frequency content of S1 as shown in Figure 4b and then to the second half to analyse the frequency content of S2 as shown in Figure4c. A 512 points FFT is applied to S1 and S2. At this stage the sound S1 or S2 cannot be separated.

In fact; the application of the FFT on heart sounds S1 and S2 after their separation or identification show that the basic frequency components are obviously detected by the Fourier transform.

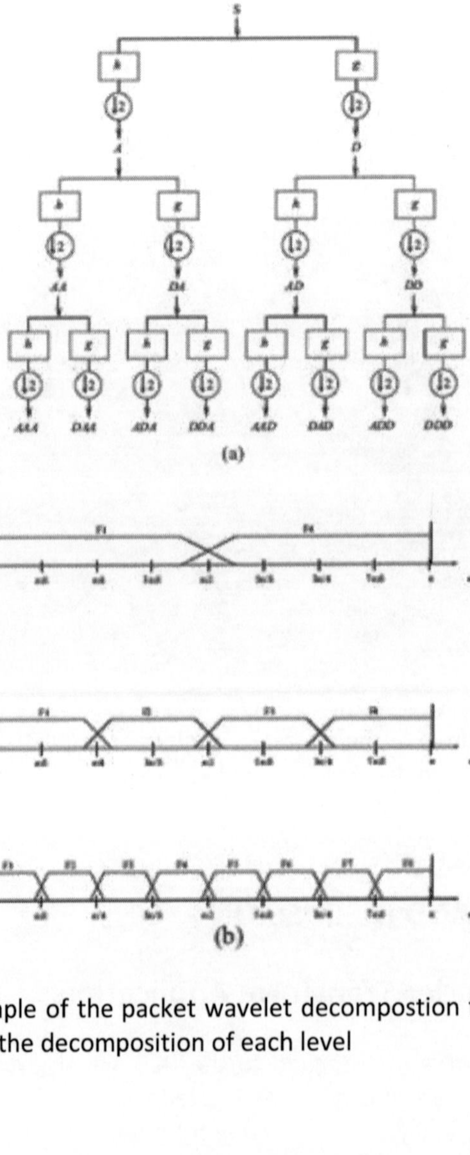

Figure 3. a) Exemple of the packet wavelet decompostion for three levels; b) Filter banks of the decomposition of each level

The spectrum of S1 has reasonable values in the range 10-180Hz. The spectrum is distinctly resolved in time into two majors components (M1 and T1) while the most of the energy of these sounds appears to be concentrated.

The two components A2 (due to the closure of the aortic valve) and P2 due to the closure of the pulmonary valve) of the second sound S2 are obvious in Figure 4c. The spectrum of the sound S2 has reasonable values in the range 50-250Hz. The spectrum for this sound is distinctly resolved in time into two majors components (A2 and P2) as shown in Figure4c.

However the FFT analysis of S2 cannot tell neither which of A2 and P2 precedes the other, nor the value of the time delay known as the "split" which separate them. For a normal heart activity usually A2 precedes P2 and the value of the split is lower than 30ms. This time delay between A2 and P2 is very important to detect some pathological cases. The sound S2 seem to have higher frequency content than that of S1 as shown in Figure4b and Figure4c.

The FFT is applied also to analyse fourth PCG signals one normal and three different marked pathological cases (the aortic-insufficiency, the aortic-stenosis and the mitral-stenosis). These are illustrated in Figure5 along with the normal PCG signal. The basic frequency content is obviously different from that of the normal PCG signal. It is clearly shown that there is great loss of frequency component in each of the pathological case with respect to normal case. In addition except the aortic- insufficiency case where we note the apparition of frequency component higher than 200Hz, the other cases (mitral-stenosis and aortic-stenosis) present a frequency spectrum limited to 200Hz.

The aortic-insufficiency and the aortic-stenosis are two pathological cases resulting from a severe organic attack, which generally involves a disappearance of the aortic component A2 of the sound S2. This shown in their corresponding PCG frequency responses illustrated in Figure5, where we notice a lack in frequency contents in the range under 100Hz compared to the normal case, where there is much more frequency component in this range. On the other hand the mitral-stenosi is rather a severe attack of the mitral valves thus involving a presystolic reinforcement as well as a bursting of the sound S1.

As the frequency extent of the sound S1 is less important than that of the sound S2, the spectral response of the PCG signal related to this pathological case is not much affected comparad to that of the normal case as was the case in the aortic -insufficiency and aortic-stenosis

In conclusion, and by applying the spectral analysis to different PCG signals, we can affirm which of the sounds S1 or S2 is directly concerned by the pathology, and more precisely which component of these sounds is affected.

With regard to normal PCG the basic frequency components are obviously detected by the FFT but not the time delay between these components. In fact as it was shown for example in Figure4c, the components A2 and P2 of the second sound S2 are obvious. However the FFT analysis of S2 cannot tell what is the value of the time delay between A2 and P2. It is thus essential to look for a transform which will describe a kind of "time-varying" spectrum.

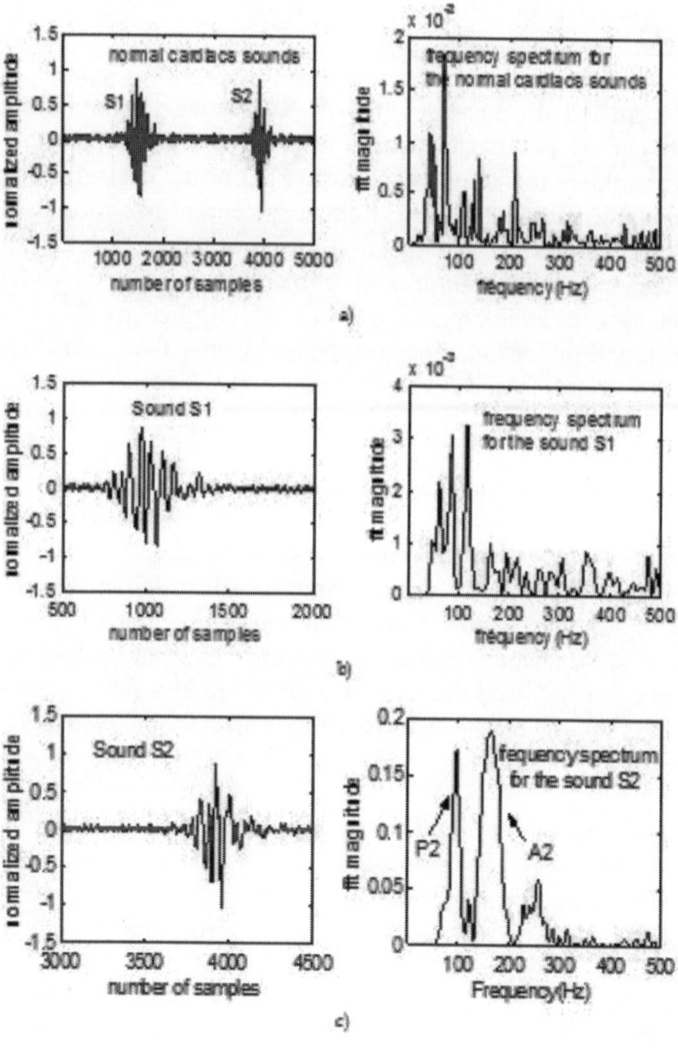

Figure 4. Frequency spectrum for the normal cardiac sounds and the sounds S1 and S2

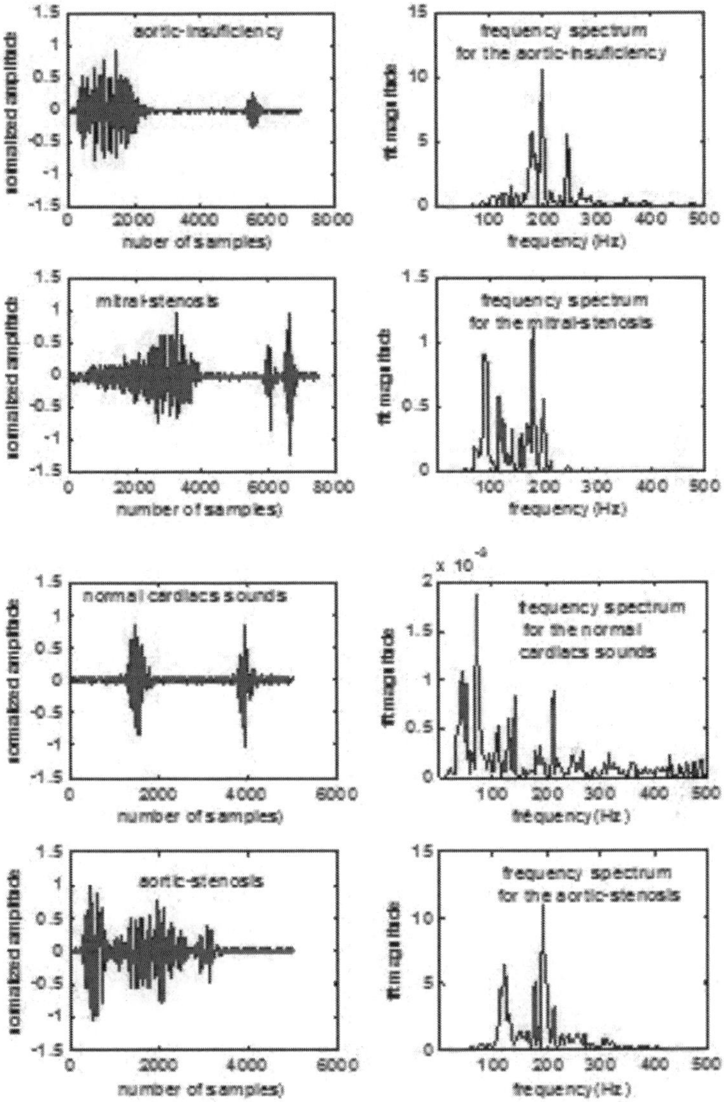

Figure 5. Normal and abnormals cardiacs sounds and their frequency spectrum respectively

3.2. Short-Time Fourier Transform Application

The normal phonocardiogram signal (Figure 6a) and the coarctation of the aorta (Figure 6a) are analyzed in this section. The coarctation of the aorta has been deliberately chosen here to evaluate the performance of the STFT analysis for it is very similar to the normal case. Figures 6aillustrate thus such a signal where we can notice that the temporal representation is almost similar to that of normal case given in Figure 6a.

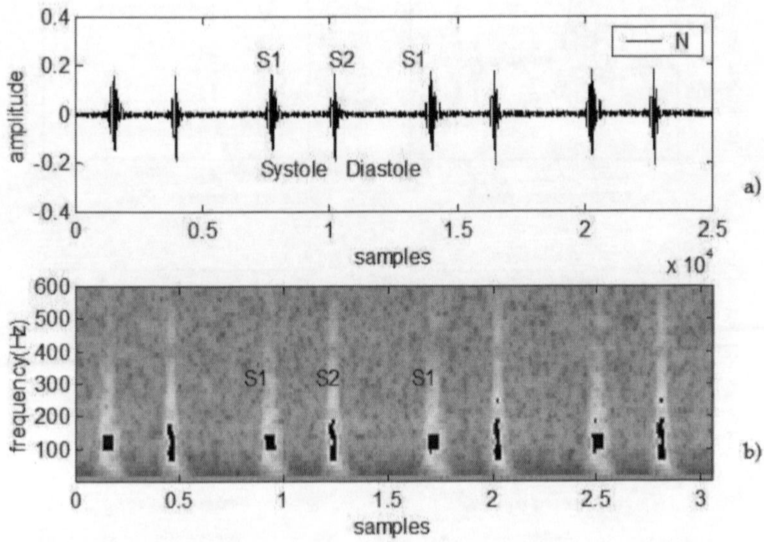

Figure 6. The STFT application of the normal PCG (N); a) the normal PCG signal (N); b) the spectrogram of the normal signal (N)

Figure 6b and Figure 7b provide respectively the STFT application of the normal and the CA case. From these Figures we can see the difference of the time-frequency features between them. For the two cases (N and CA), the second sound (S2) is shown to have higher frequency content than of the first sound (S1) [14,15]. This expected since the amount of blood present in the cardiac chambers is smaller [16].

We consider here two examples of the phonocardiogram signals with murmur: the pulmonary stenosis (PS :Figure 8a) with a systolic murmur and the aortic regurgitation (AR : Figure 9a) with a diastolic murmur.

Figure 8b and Figure 9b shows respectively the STFT application of the PS and AR signals. We can notice that the frequency extent of the diastolic murmur of the AR case is highly (About 600Hz) than the systolic murmur of the PS case (about 400 Hz).

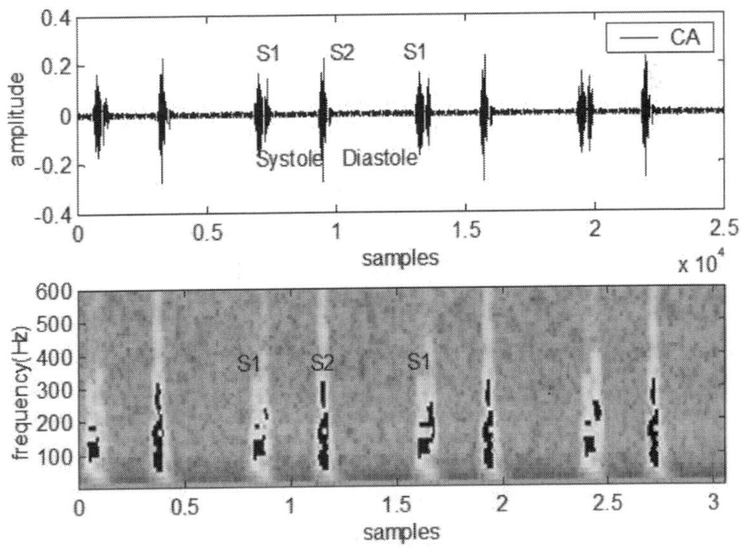

Figure 7. The STFT application of the Coarctation of the aorta (CA); a) the Coarctation of the aorta signal (CA); b) the spectrogram of the signal CA

In this section we presents the experimental results of the short-time frequency transform application of the three followings groups of the PCG signals used.

- Group 1 : normal (N) or similar morphological signal (CA) ;
- Group 2 : opening snap (OS) and ejection click (EC);

- Group 3: PCG signal with width murmur (PS and AR).

Figure 10 provides a better representation of the results obtained concerning the frequency contents of the sounds and murmurs analysed. If under the normal conditions (N) or in the presence of similar signals (CA) the frequency content of the sound S2 is more significant than that of the sound S1.

We noted that the light murmurs (OS, EC.) can influence the time-frequency content of the principal sounds S1 and S2 and have a frequency extent more significant than them.

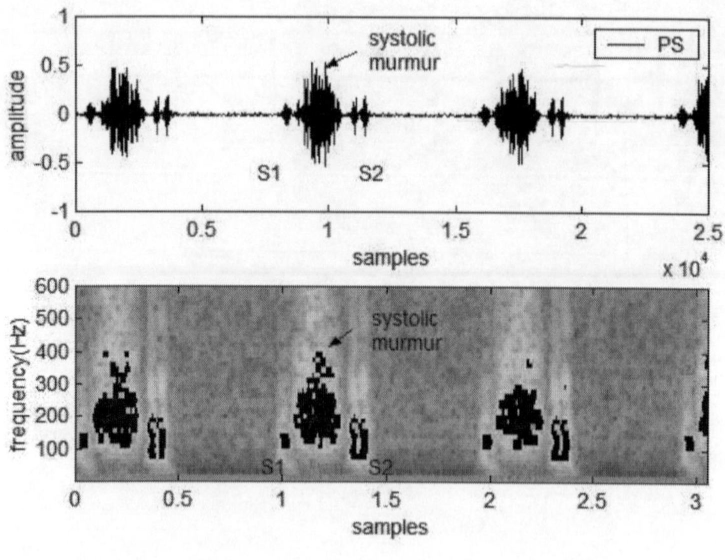

Figure 8. The STFT application of the pulmonary stenosis; a) The pulmonary stenosis signal (PS); b) The spectrogram of the signal PS

Finally in fact the width murmurs (PS and AR cases) present a frequency extent very significant. Discrimination between the systolic and diastolic murmurs can be made starting from this frequency extent, diastolic murmurs thus having a frequency extent more significant than the systolic murmurs. In more these murmurs seem not too much not to affect the time-frequency content of the sounds S1 and S2.

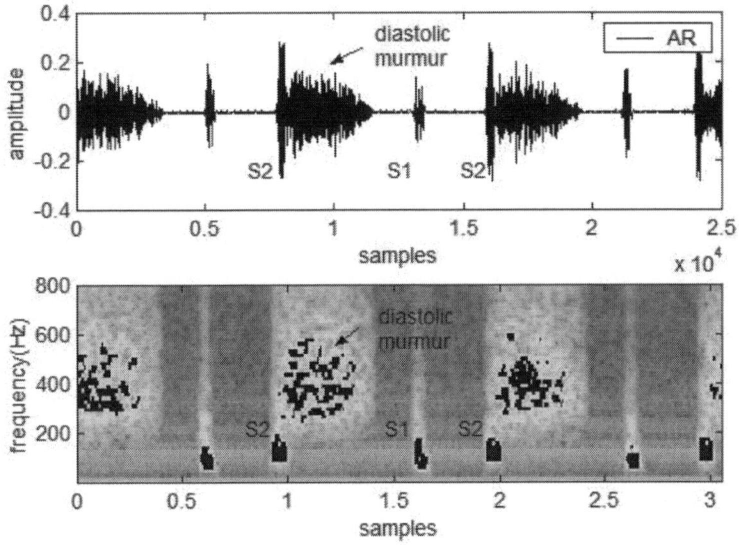

Figure 9. The STFT application of the aortic of the regurgitation; a) The aortic of the regurgitation signal (AR); b)The spectrogram of the signal AR

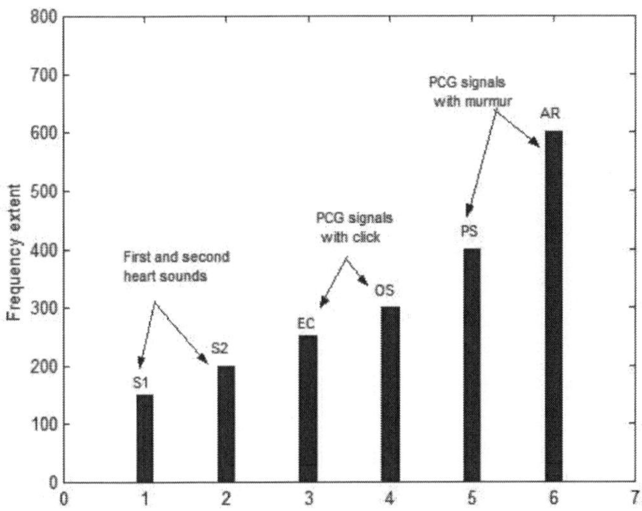

Figure 10. Frequency extent of the three groups of the PCG signals

3.3. Wigner Distribution Application

Figure11 shows the WD application of the cardiac sound of the normal case (11a), the coarctation of the aorta (11b) and the innocent murmur (11c). One can notice here that the two principal components (A2 and P2) start to appear in the presence cross-terms. The WD results may be improved by increasing the sampling rate of original signal,, but it still suffers from the cross-terms problem because of the nonlinearity of the WD.

However the Wigner distribution (WD) have shown good performances in the analysis of non-stationary signals. This comes from the ability of the WD to separate signals in both time and frequency directions. One advantage of the WD over the STFT is that it does not suffer from the time-frequency trade-off problem. On the other hand, *the WD has a disadvantage since it shows cross-terms in its response. These cross-terms are due to the nonlinear behaviour of the WD, and bear no physical meaning*. One way to remove these cross-terms is by smoothing the time-frequency plane, but this will be at the expense of decreased resolution in both time and frequency.

3.4. Continuous Wavelet Transform Application

An algorithm under MATLAB environnement of the Continuous Wavelet Transform is developed then applied to analyse the different PCG signals. First the normal PCG is studied as illustrated byFigure12.a. The two heart sounds are clearly shown in dark color. They are spaced with 2500 samples corresponding to 0.312 secondes. The continuous wavelet transforms of S1 and S2 are also displayed separately in Figure12.b and Figure12.c respectively. The sound S2 is shown to have higher frequency content than that of the S1. This is expected since the amount of blood present in the cardiac chambers is smaller [1],[16]. On Figure 12b we can see that S1 is clearly resolved in two major component (M1 and T1). On the figure 10c the sound S2 is resolved also into two major's components (A2 and P2). The time delay between A2 and P2 can be easily measured with the use of the wavelet coefficients (Figure12c). This delay is measured to be 13ms. It is smaller than the 30ms [17] as foreseen in the normal conditions of the PCG signal. Pathological conditions could cause this time difference to narrow or widen. Moreover, the order of occurrence of A2 and P2 may be reversed. The wavelet transform allows measurement and determination of this time difference, and thus allows a diagnosis process regarding this important parameter to be produced. Table I resumes the differences observed between the components A2 and P2.

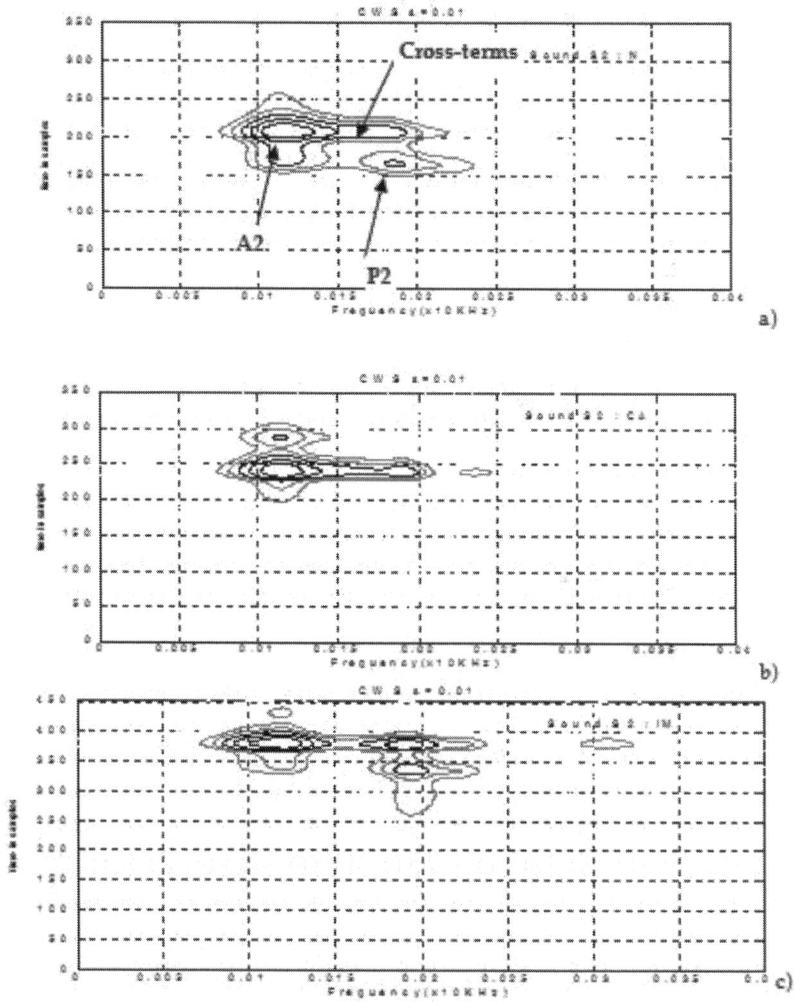

Figure 11. The wigner distribution of the second cardiac sound S2 : a) the normal case, b) the coarctation of the aorta case (CA), c) the innocent murmur case (IM)

Table 1. Temporal and frequential measurements related to the componentsA2 and P2.

	Localisation of the Delay "d" (ms)	Minimal frequency (in scale)	Maximal frequency (in scale)	Frequency Extent
A2	13	19	124	105
P2	13	18	116	98

It can be concluded for the normal PCG that :

1) The component A2 precedes in time the component P2.

2) The component A2 have higher frequency content than P2.

3) The amplitude of A2 is more important than that of P2.

These parameters, particularly the frequency, make it possible to see a difference between A2 and P2.

Moreover the ability of the wavelet transform in heart disease diagnosis can be studied by applying the CWT algorithm on different marked cases. The result of this application are illustrated in Figure13b(aortic-insufficiency), Figure13c (aortic-stenosis) Figure13d (mitral-stenosis). The coefficients of the CWT allow us to clearly discern the frequency range of each signal. It also shows the major components according to the temporary variation; the maximal amplitude is characterised by a darker color than those of the small amplitudes.

3.5. Discrete Wavelet Transform Application

The multiresolution analysis based on the discrete wavelet transform (DWT) is a powerful tool in and filtering, separating and identification of the internal components and murmurs of the various analyzed signals (Figure14).

Figure15a shows the application of the discrete wavelet transform of one cycle of the normal PCG. Levels d1 and d2 represent the high frequency variations of the base line of PCG signal. Levels d3 and d4 emphasize clearly the side of high frequency content of the sound S2 compared to S1The component A2 (the most predominant in sound S2) appears better on levels d5 and d6. Level d7 represents the two principal components of the sound S1 (M1, T1).

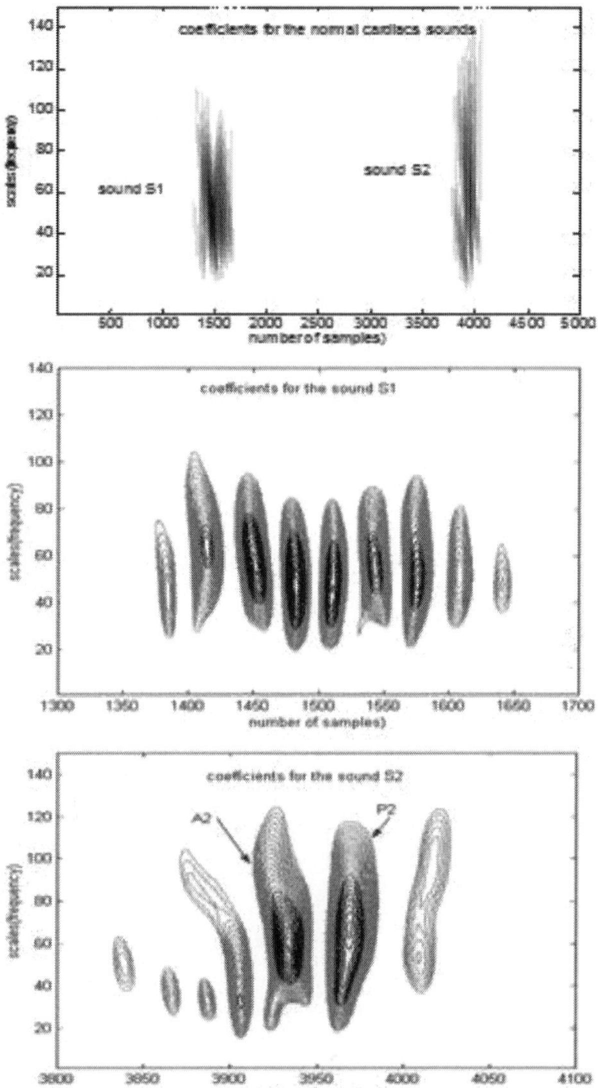

Figure 12. Continuous wavelet transform analysis (CWT) for the normal cardiac sounds S1 and S2

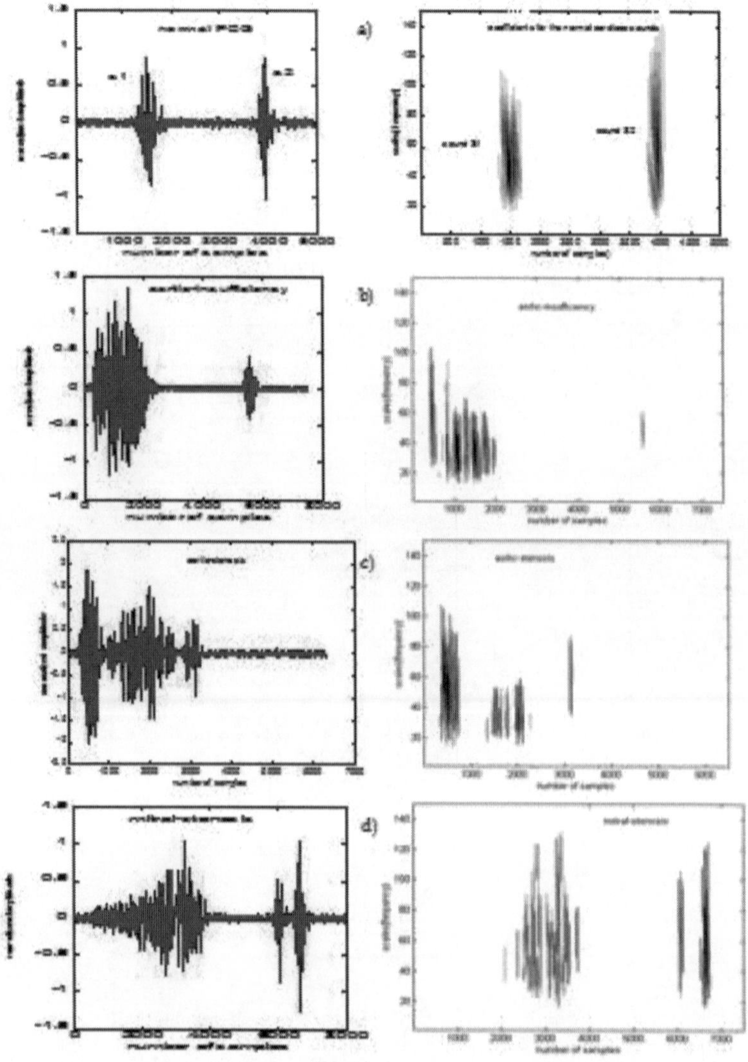

Figure 13. Continuous wavelet analysis (CWT) for the normal PCG and abnormal PCG.

The same analysis by using the discrete wavelet transform concerning the coarctation of the aorta and the mitral stenosis is also applied (Figures16b and Figure16c). Figure 13b, in level d7, provide a well representation of the third component "C" added a the two majors components of the sound S2 (A2 and P2).

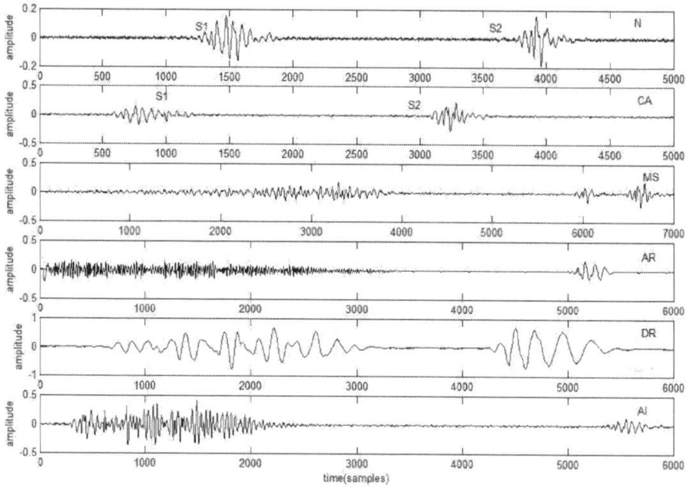

Figure 14. PCG signals used a) normùal (N), b) the coarctation of the aorta (CA); c) the mitral stenosis (MS), d) the aortic regurgitation (AR), e) the diastolic ruble (DR), f) the aortic insufficiency (AI)-

3.6. Packet Wavelet Transform Application

The used of the wavelet packet transform is exactly the same as those developed in the discrete wavelet transform. The only difference is that wavelet packets offer a more complex and flexible analysis, because in wavelet packet analysis, the details as well as the approximations are split (Figure3). Single wavelet packet decomposition gives a lot of bases from which you can look for the best representation with respect to a design objective. The *wavelet packet* method is thus a generalization of wavelet decomposition that offers a richer range of possibilities for signal analysis. In wavelet analysis, a signal is split into an approximation and a detail. The approximation is then itself split into a second-level approximation and detail, and the process is repeated.

The wavelet packet transform analysis in this paper give important features of the extent frequency of the heart sounds (S1 or S2) and cardiac murmur.

These features can be help clinician in their diagnosis or recognizing pathological conditions concerning the recording PCG signals.

Figure 16 provide a time frequency (TF) representation of one cardiac cycle of the heart sounds concerning the normal PCG signal (Figure 16a), the coarctation of the aorta case (Figure 16b) and the mitral stenosis case (Figure 16c). These figures shown the frequency range of each component or murmur of the PCG signal studied. Thus we can observe the component "C" added of the two majors component for the sound S2 (A2 and P2). The same result have been find in the section 3.5 (Figure 14b. detail d7)

Figure 17 has the advantage of presenting at the same time the frequency extent of the various components of the cardiac sound like their frequency site one compared to the other. We can shown clearly that the Diastolic ruble (DR) case have high frequency content that the aortic insufficiency case (AI) or the aortic regurgitation case (AR).

4. CONCLUSION

The cardiac (heartbeat sound) cycle of phonocardiogram (PCG) is characterized by transients and fast changes in frequency as time progresses. It was shown that basic frequency content of PCG signal can be easily provided using FFT technique. However, time duration and transient variation cannot be resolved; the CWT wavelet transform therefore i is a suitable technique to analyse such a signal. It was also shown that the coefficients of the continuous wavelet transform give a graphic representation that provides a quantitative analysis simultaneously in time and frequency. It is therefore very helpful in extracting clinically useful information.

The measurement of the time difference between the A2 and P2 components in the sound S2, the number of major components of the sounds S1 and S2 and the frequency range and duration for all these components and sounds can be accurately achieved for the CWT simultaneously as was clearly illustrated.

It is found that the wavelets transform is capable of detecting the four major components of the first sound S1 and the two components (the aortic valve component A2 and the pulmonary valve component P2) of the second sound S2 of a normal PCG signal. These components are not accurately detectable using the STFT or WD. However the standard FFT can display the frequencies of the components A2 and P2 but cannot display the time delay between them.

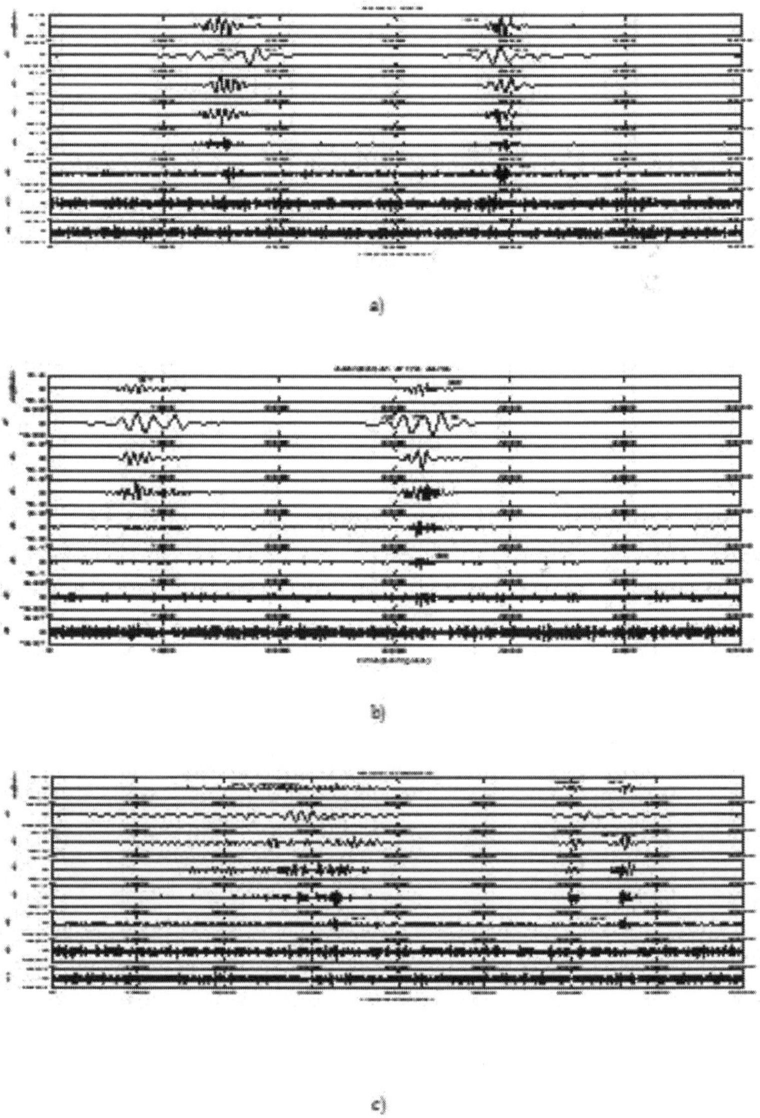

Figure 15. Discrete wavelet transform (DWT) analysis for: a) the normal PCG, b) the Coarctation of the aorta, c) the mitral stenosis

The application of the STFT in the analysis of the PCG signals made it possible to obtain appreciable information on the time-frequency content of the sounds S1, S2 and of the added murmurs (OS,EC or width murmurs).

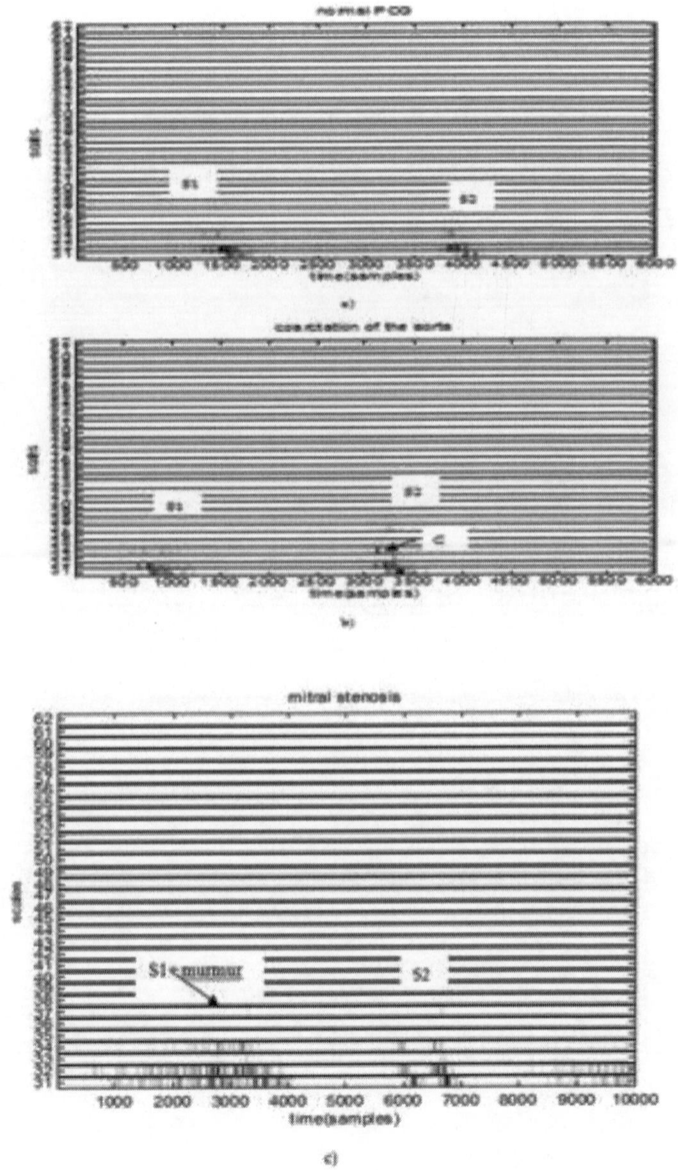

Figure 16. Wavelet packet transform (WPT) analysis for: a) the normal PCG, b) the coarctation of the aorta, c) the mitral stenosis

If under the normal conditions (N) or in the presence of similar signals (CA) the frequency content of the sound S2 is more significant than that of the sound S1. We noted that the light murmurs (OS, EC.) can influence the time-frequency content of the principal sounds S1 and S2 and have a frequency extent more significant than them.

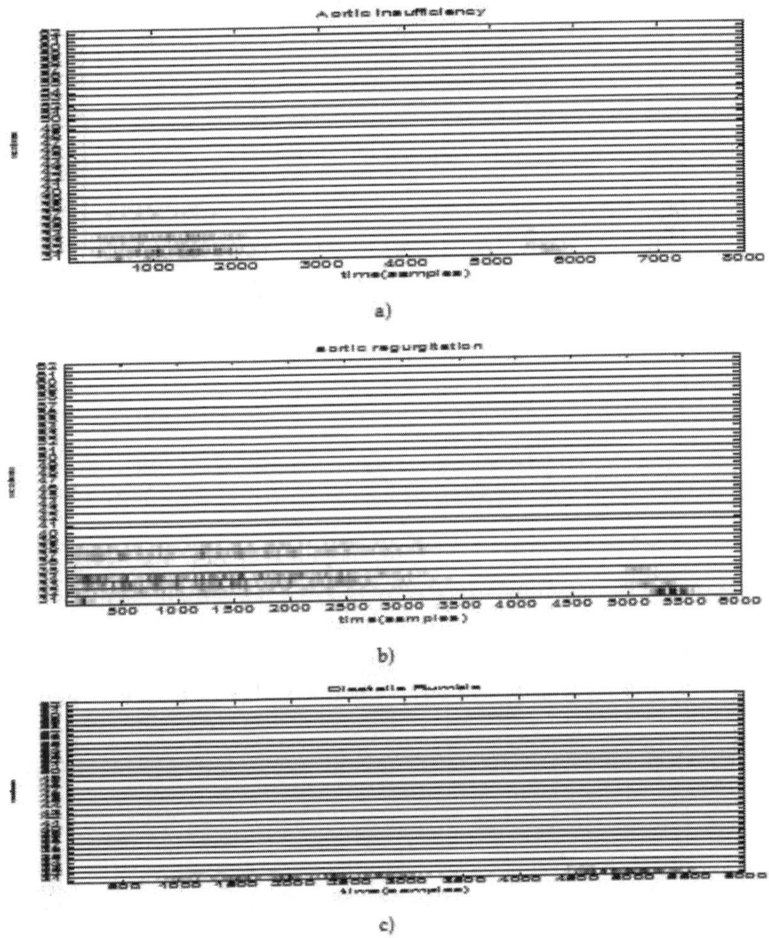

Figure 17. Wavelet packet Transform (WPT) analysis for: a) the aortic Insufficiency (Ai), b) the aortic regurgitation (AR), c) the Diastolic Rumble (DR)

Finally in fact the width murmurs (PS and AR cases) present a frequency extent very significant. Discrimination between the systolic and diastolic murmurs can be made starting from this frequency extent, diastolic murmurs thus having a frequency extent more significant than the systolic murmurs. In more these murmurs seem not too much not to affect the time-frequency content of the sounds S1 and S2.

The two version of analysis of the wavelet transform (DWT and PWT) make it possible to gather time-frequency information concerning the characteristics of the cardiac sounds.

It is shown that the FFFT, the STFT, the WD and the WT techniques provides more information of the PCG signals with murmurs that will help physicians to obtain qualitative and quantitative measurements of the time and the time-frequency PCG signal characteristics and consequently aid to diagnosis.

REFERENCES

1. Rangayyan,R.M and Lehner,R.J 1988A review.CRC Critical Reviews in Biomedical Engineering 15 (3), 211-236.

2. Rangayyan,R.M and Lehner,R.J 1988Phonocardiogram signal analysis: a review. CRC Critical Reviews in Biomedical Engineering 15 (3), 211-236.

3. Feigen, L.P 1971Physical characteristics of sound and hearing. American journal of Cardiology, 28 (2), 130-133.

4. Tuteur, F.B 1988Wavelet Transforms in signal detection. IEEE ICASSP, CH2561-9, 1435-1438.

5. Grossmann.A; Holschneider Kronland-Martinet.R and Morlet,J 1987Detection of abrupt Changes in sound signal with the help of the wavelet transform. In: Inverse problemes: An interdisciplinary study. Advances in Electronics and Electron physics. Supplement 19 [New York, Academic], 298 306

6. Obaidat.M.S. Phonocardiogram signal analysis: techniques and performance comparison.Journal of Medical Engineering & technologie, 17 6November-December 1993

7. B. Boashas, 1993Time-frequency signal analysis. In advances in spectrum Estimation, edited by S.Haykin, (NJ: Prentice-Hall).

8. William J.Williams 1997Time-frequency and wavelets in Biomedical Signal Processing. Edited by Metin Akay. IEEE Press Serie in BME. 3 8

9. Olivier.R and Duhamel.P.Fast algorithms for Discrete and Continuous Wavelet Trnsforms. IEEE Transactions Information Theory, 38 2march 1992

10. A. Harris, G. Sutton, Towers. M. And, 1976Physiological and clinical aspects of cardiac auscultation. Medicine Ltd, London, Uk.

11. Y.Meyer 1990Ondelettes et Opérateurs. Tome1. Paris: Hermann.

12. S.Mallat 1989The wavelet representation. IEEE Trans.Pattern Anal.Machine Intell, vol 11, A theory for multiresolution signals decomposition: 674 693

13. P.I.J. Keeton, F.S. Schlindwein, Application of Wavelets in Doppler Ultrasound, MCB University Press, Emerald Sensor Review, Vol. 1997 38 45

14. S.M Debbal and F.Bereksi-Reguig.The Fast Fourier and the wavelet transforms analysis of the cardiacs sound. Physical and Chemical News (PCN). 15 54 59 2004

15. S. M. Debbal-F, Reguig. «. Bereksi, fast. The, transform. Fourier, continuous. the, transform. wavelet, of. analysis, phonocardiogram. the, signal, of. Journal, in. Mechanics, Medicine, . J. M. M. B. Biology, I. S. S. N. 0219-519, vol., . No, September, 2004 257 272

16. P. M. Bentley, A. Grant, J. T. E. Mc Donnel, Time-frequency and Time-scale for the classification of native and Bioprosthetic valve sounds. IEEE Trans. Biomed. Eng. 45(1), 125 128 1998

17. Djebbari.A 1999Synthèse des méthodes d'analyse temporelle, spectrale et spectro-temporelle du signal phonocardiogramme. Electronic magister thesis of signals andsystems. Departement of electronics, faculty of science engineerig, university Aboubekr belkaid Tlemcen, Algeria, 18 24

Chapter 5

A wavelet-based super-resolution method for multi-slice MRI

Rafiqul Islam[1], Andrew J. Lambert[1], Mark R. Pickering[1], Jennie M. Scarvell[2], Paul N. Smith[2]

[1]School of Engineering and Information Technology (SEIT), University of New South Wales, Canberra, Australia

[2]The Trauma and Orthopaedic Research Unit, The Canberra Hospital, Canberra, Australia

ABSTRACT

In multi-slice magnetic resonance imaging (MRI), the resolution in the slice direction is usually reduced to allow faster acquisition times and to reduce the amount of noise in each 2D slice. To address this issue, a number of super resolution (SR) methods have been proposed to improve the resolution of 3D MRI volumes. Most of the methods involve the use of prior models of the MRI data as regularization terms in an ill-conditioned inverse problem. The use of user-defined parameters produces better results for these approaches but an inappropriate choice may reduce the overall performance of the algorithm. In this paper, we present a wavelet domain SR method which uses a Gaussian scale mixture (GSM) model in a sparseness constraint to regularize the ill-posed SR inverse problem. The proposed approach also makes use of an extension of the Dual Tree Complex Wavelet Transform to provide the ability to analyze the wavelet coefficients with sub-level precision. Our results show that the 3D MRI volumes reconstructed using this approach have quality superior to volumes produced by the best previously proposed approaches.

KEYWORDS

Magnetic Resonance Imaging; Super Resolution; Gaussian Scale Mixture Model; Wavelet Regularization

1. INTRODUCTION

Magnetic resonance imaging (MRI) is an imaging technique for producing images of parts of the human body for clinical purposes. When compared to other widely used scanning methods such as Computed Tomography (CT) and Positron Emission Tomography (PET), MRI has the advantage of using only strong magnetic fields and non-ionizing radiation in the radio frequency range. In multi-slice MRI, multiple 2D slices are captured to form a 3D volume of the desired body part. Recently, faster multi-slice acquisition methods (such as Single-shot Spin Echo, Conventional Spin Echo, etc.) are often used to capture MRI slices in the presence of motion of the imaging subjects [1,2]. During this faster acquisition, high-resolution image slices are obtained in the in-plane direction but some undesirable artifacts are observed due to the resolution reduction in the slice-select direction [3]. As a result of this resolution reduction, some valuable information may be lost [4]. To deal with this problem, a number of image super resolution (SR) techniques have been used as post-processing methods to improve the quality of the low-resolution images [5].

The basic idea underlying all SR algorithms is the combination of a number of low-resolution (LR) blurred noisy images to produce an SR image [6]. Within this wide area of study, medical image super resolution has emerged as a particularly active field and initial attempts at the application of SR reconstruction algorithms for MRI images were made in [7,8]. In [7], the authors utilized an iterative back projection (IBP) method [9] to improve the in-plane resolution of diffusion-weighted images. In their method, the low resolution images were acquired by sub-pixel shifting in the phase-encode and frequency-encode direction. Since each of the sub-pixel shifted images contained identical information but with different noise [10,11] a small improvement in resolution can be obtained. In [8] the authors improved this method by utilizing sub-pixel shifts in the slice-select direction instead of in the in-plane direction during the multi-slice acquisition. In their work, a number of multi-slice low resolution volumes were acquired where each volume was shifted with respect to the other volumes by a subpixel amount in the slice-select direction. These multiple parallel overlapping slice stacks were combined using their proposed super resolution reconstruction algorithm to produce a 3D isotropic volume with improved resolution and sharper edges in the slice-select direction. The authors in [12,13] also utilized the IBP reconstruction method to improve the resolution in the sliceselect direction. However, the blur and sharpening kernel used in their approach introduced some unwanted noise and artifacts. The SR method proposed in [13] was more effective in improving the slice-select direction resolution as a more appropriate blur kernel was utilized.

More recently, the use of regularization prior models has improved the performance of SR algorithms. An SR reconstruction technique based on the use of a discontinuity-preserving regularization method was proposed in [14] The authors demonstrated that an SR approach using this priori model increased the ability of clinicians to detect and visualize small regions of neuronal activity. Moreover, the activated regions appeared sharper and provided better information regarding their morphological limits and structure. Shantanu et al. [15] proposed an MRI resolution enhancement algorithm that made use of a total variation (TV) prior. They also utilized an edgepreserving operator during the up-sampling and downsampling of images that enhanced the spatial resolution. Recently, a spatial-domain regularization approach for SR was proposed in [16]. This regularization method, known as directional bilateral total variation (DBTV), applied a directionally adaptive weighting on the traditional total variation regularization approach. Schilling et al. introduced a multi-stack approach in [17] where a number of LR MRI volumes were captured at different orientations and combined to produce an SR volume using the simultaneous additive reconstruction technique (SART). They showed that capturing the LR images at different orientations produced superior results than when using a parallel stack approach [8] where the LR images were captured at different distances along the slice-select direction only.

In this paper, we address the problem of choosing an appropriate regularization prior model in SR reconstruction methods to improve the resolution of MRI volumes in the slice-select direction. In [18], we demonstrated the benefits of using a GSM model of a standard wavelet transform as a prior for simulated MRI data. In the approach presented here we adopt an extended waveletdomain regularization prior model to analyze the wavelet coefficients with sub-level precision. Our experimental results for both simulated and real MRI data show that this algorithm produces SR volumes with quality superior to volumes produced by the best previously proposed approaches.

The remainder of the paper is organized as follows: In Section 2, we explain the concept of our sub-level wavelet transform. The general approaches used for SR are described in Section 3 and our proposed extensions to this approach for generating a super-resolution volume are presented in Section 4. Section 5 is devoted to describing the experimental evaluation of the algorithms and finally our conclusions are presented in Section 6.

2. THE SUB-LEVEL WAVELET TRANSFORM

The wavelet transform has been widely used as a multiresolution data analysis method across the field of signal processing for applications such as image denoising, image de-blurring and image enhancement. The DTCWT, originally introduced by Kingsbury in [19] has been shown to be a superior wavelet transform in the field of image processing because of its shift invariant and directionally oriented properties. In general, the wavelet transform of an image x can be expressed as follows:

$$\psi_j = \mathbf{W}_j(\mathbf{x}) \tag{1}$$

where W_j represents the wavelet transform operation and ψ_j are the wavelet coefficients for scale levels j = 1, 2 ××× L.

In this paper, we extend the DTCWT by adding an extra level in between each traditional level. This extra level provides more information during the subsequent analysis of the images. The new wavelet transform of an image x is implemented by adding an extra level and the overall operation of this transformation can be described as follows: Firstly the normal DTCWT is calculated as in Eq.1. Then to add an extra level, we resize the original volume by a factor of $2^{-0.5}$ using linear interpolation and apply the wavelet transform again. This transformation produces wavelet coefficients for scale levels j = 1.5, 2.5 ××× L + 0.5. Finally, we merge these two sets of wavelet coefficients to produce the new wavelet coefficients for scale levels j = 1, 1.5, 2 ××× L, L + 0.5. These extended set of coefficients effectively have sub-level resolution and provide the ability to better analyze the inter-level relationships between coefficients during the component wise de-noising stage of our proposed SR framework. The overall sub-level wavelet decomposition process using the above process is illustrated in **Figure 1**.

3. BACKGROUND

The main objective of image deconvolution algorithms is to find the solution of an ill-conditioned equation of the form:

$$\mathbf{y} = \mathbf{H}\mathbf{x}_0 + \mathbf{b} \tag{2}$$

where x_0 is a vector of pixel values from the original image, y is a vector of pixel values from the observed blurry image, H represents a convolution (i.e. blockcirculant) matrix that approximates the blurring function and b is a vector representing white noise with zero mean and variance σ^2[20].

A solution to this minimization problem can be explained by first considering the case when the noise term b is set to zero. The cost function for this case is given by:

$$J(\mathbf{x}) = \|\mathbf{y} - \mathbf{H}\mathbf{x}\|_2^2 \tag{3}$$

The solution that minimizes this cost function is given by the gradient descent method as follows:

$$\mathbf{x}^{(n+1)} = \mathbf{x}^{(n)} + \tau\mathbf{H}^T\left(\mathbf{y} - \mathbf{H}\mathbf{x}^{(n)}\right) \tag{4}$$

where τ is a fixed step size and H^T represents the backprojection kernel that should be the pseudo-inverse of the convolution matrix. This equation represents the classic Landweber iteration [21].

Daubechies et al. proposed a solution to the problem of estimating the value of x_0 when both blurring and noise are present by using the thresholded-Landweber algorithm [22]. This approach consists of first performing the Landweber iteration with step-size τ:

$$\mathbf{z}^{(n)} = \mathbf{x}^{(n)} + \tau \mathbf{H}^T \left(\mathbf{y} - \mathbf{H}\mathbf{x}^{(n)} \right) \tag{5}$$

and then performing the wavelet domain de-noising operation:

$$\mathbf{x}^{(n-1)} = \mathbf{M}\phi\left(\mathbf{W}\mathbf{z}^{(n)}, \frac{\tau\lambda}{2} \right) \tag{6}$$

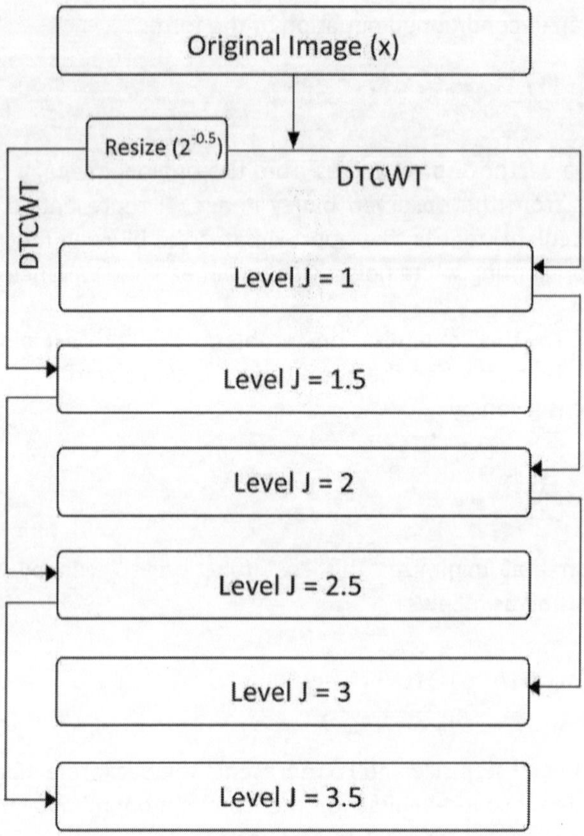

Figure 1. Flowchart for a 3 level wavelet decomposition with an extra tree based on a resized image.

The function f(w,t) is the soft-thresholding function defined in [22,23] with the form:

$$\phi(w,t) = \mathrm{sgn}(\mathrm{w})(|w| - t)_+ \tag{7}$$

where $(.)_+$ is the positive-part function given by:

$$(y)_+ = \begin{cases} y & \text{if } y > 0 \\ 0 & \text{otherwise} \end{cases} \tag{8}$$

Vonesch and Unser [23] improved this approach by developing the fast thresholded-Landweber (FTL) deblurring algorithm. In their approach, the original thresholded-Landweber operation is performed on a wavelet sub-band basis using different step-sizes for each subband as follows: For each sub-band j, perform the Landweber iteration:

$$\mathbf{z}_j^{(n)} = \mathbf{W}\left(\mathbf{x}^{(n)} + \tau_j \mathbf{H}^T \left(\mathbf{y} - \mathbf{H}\mathbf{x}^{(n)}\right)\right) \tag{9}$$

where W_j represents the wavelet transform operation for sub-band j, $\tau_j = 1/\alpha_j$ and α_j is the sub-band emphasis factor defined in [23]. Then perform the wavelet domain soft-thresholding de-noising operation on each sub-band and apply the inverse wavelet transform to find the updated version of x as follows:

$$\mathbf{x}^{(n+1)} = \sum_{j \in S} \mathbf{M}_j \phi\left(\mathbf{z}_j^{(n)}, \frac{\tau_j \lambda}{2}\right) \tag{10}$$

where M_j represents the inverse wavelet transform operation for sub-band j.

Zhang and Kingsbury [24] extended this algorithm by utilizing the element wise de-noising operation on each sub-band instead of the soft-thresholding de-noising function. The de-noising operation for every element in the sub-band j is given by

$$w_i^{(n+1)} = \frac{\left(\hat{w}_i + \varepsilon^2\right)}{\alpha_j\left(\hat{w}_i^2 + \varepsilon^2\right) + \sigma^2} z_i^{(n)} \tag{11}$$

where w_j and z_j are coefficients in the sub-band j, w_t is calculated using the bivariate shrinkage de-noising algorithm defined in [25] on the wavelet coefficients of $x^{(n)}$, σ^2 is the variance of measurement noise and ε is a stabilizer to avoid infinity. An inverse wavelet transform is then used to find the new estimate.

$$\mathbf{x}^{(n+1)} = \sum_{j \in S} \mathbf{M}_j \mathbf{w}_j^{(n+1)} \tag{12}$$

This de-noising approach was shown by Zhang and Kingsbury in [24] to be equivalent to replacing the l_1 norm sparseness constraint with a sparseness constraint that is closer to the l_0 norm.

4. SUPER RESOLUTION USING A GSM MODEL CONSTRAINT

The resolution of any image is dependent on the properties of the sensors in the imaging device. Hence, the objective of a super resolution algorithm is to find the solution of an ill-conditioned equation of the form:

$$\mathbf{y}_i = \mathbf{DHF}_i \mathbf{x}_0 + \mathbf{b} \tag{13}$$

where y_i is a vector of pixel values from the ith observed image, F_i is a warping matrix, D is a sub-sampling matrix and b_i represents the noise.

Since estimating x_0 given y_i is an ill-posed inverse problem, it is necessary to rely on prior information about the original signal to obtain an accurate estimate with robustness to noise. Including this prior information results in a minimization problem where the quality of a given estimate x is measured by a cost function of the form:

$$J(\mathbf{x}) = \left\| \mathbf{y}_i - \mathbf{DHF}_i \mathbf{x} \right\|_2^2 + \lambda \left\| R(\mathbf{x}) \right\|_1 \tag{14}$$

The first term in this cost function quantifies the error between the observed LR image y_i and an estimate of y_i calculated from the current estimated high resolution image x. The second term is designed to penalize the estimated image when it does not conform to the prior model and thus helps to obtain a stable solution. The regularization parameter λ balances the contribution of both terms. The super resolution problem now becomes one of finding an estimate x of the original higher resolution image that minimizes the cost function J(x).

In this paper, for the regularization prior model, we use the wavelet domain GSM model as well as the sublevel wavelet domain GSM model. The main benefits of GSM model is it has the ability to regularize the estimated MRI volumes in terms of both the heavy tail distributions and the inter-scale properties of wavelet coefficients [26,27].

In our proposed wavelet domain GSM method, we extend the modified FTL algorithm to include multiple observed images and sub-sampling in the image acquisition model. To accommodate this extension, we modify the Landweber iteration as follows:

$$\mathbf{z}_j^{(n)} = \mathbf{W}_j\left(\alpha_j \mathbf{x}^{(n)} + \sum_{i=1}^{k} \mathbf{F}_i^{-1}\mathbf{H}^T\mathbf{D}^T\left(\mathbf{y}_i - \mathbf{D}\mathbf{H}\mathbf{F}_i\mathbf{x}^{(n)}\right)\right) \quad (15)$$

where $\mathbf{F}_i^{-1}$ is the inverse of the warping operator and D^T is an up-sampling matrix. A component-wise wavelet based de-noising operation is then performed on every sub-band j and the inverse wavelet transform is applied to find the updated version of x using Eq.12.

For our proposed sub-level wavelet domain GSM method, we use the extension of the DTCWT instead of the normal DTCWT discussed in Section 2 as a wavelet transform. The main benefit of this model over the normal GSM model is that it provides the ability to analyze the inter-level relationships between sub-level wavelet coefficients with finer resolution.

The algorithm starts with an initial guess of the highresolution image x from a set of low-resolution images. In our work, we choose one reference volume and used the cubic interpolation approach to create the initial highresolution image. Then an iterative process is used by alternating between the new modified Landweber iteration and de-noising method to estimate the output. The overall procedure is summarized in **Figure 2**.

5. PERFORMANCE EVALUATION

Experiments were conducted to evaluate the performance of the proposed SR algorithm for the purposes of increasing the resolution of 3D MRI volumes. We compared the performance of our proposed algorithm with a SR algorithm based on DBTV [16] as well as the SART approach [17] as these were found to be the best performing of the recently proposed SR approaches. The wavelet transform we have used in our proposed algorithm is a 3D version of the DTCWT. In the MRI acquisition process, the point spread function (PSF) is defined by the slice excitation profile function. In our experiment, we assume the slice profile function is approximated by a Gaussian function with the full-width half maximum (FWHM) distance set to the selected slice thickness.

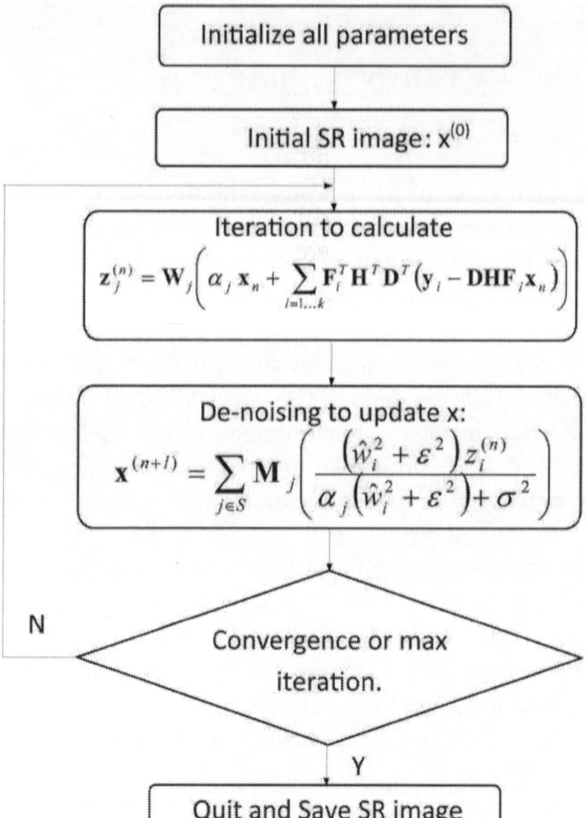

Figure 2. Flowchart for our proposed SR algorithm.

5.1. Volume Reconstruction Using Simulated Data

The input data used was a 3D MRI volume from the Brain Web Simulated Database [28]. **Figure 3**(a) shows a slice from the original volume. In our experiment, we took a 3D volume with 1 mm voxel resolution in the slice-select direction. We simulated the low-resolution volumes by rotating the original volume about all three axes simultaneously by 1, 2, 3 and 4 degrees. Each of these rotated volumes was then sub-sampled by a factor of 4 in the slice-select direction. Blur and noise were also added to simulate the effect of the image acquisition process. We used the multi-stack approach of capturing the LR volume at different orientations as this technique was shown in [17] to give better performance than the parallel stack approach. **Figure 3**(b) shows a slice from one of the sub-sampled, blurred and noisy input volumes, **Figure 3**(c) shows one slice from the output volume of the SR with DBTV algorithm, **Figure 3**(d) shows one slice from the output volume of the SR algorithm using SART, **Figure 3**(e) shows one slice from the output volume of our proposed SR algorithm using a GSM model prior [18] and **Figure 3**(f) shows one slice from the output volume of our proposed SR algorithm which incorporates both a GSM model prior and the sub-level wavelet transform described in Section 4. These images clearly show the improved subjective quality of the output of our proposed approach. The arrows in **Figure 3**(f) indicate example areas of improved resolution for the proposed algorithm.

For a practical measure of the quality of the estimated super resolution volume, we use a measure of the accuracy of a 3D segmented region produced using the estimated volumes. We call this measure the segmentation difference (SD) and it is defined as the percentage of voxels which are segmented differently for the estimated volume compared to the original volume. We use the formulae below to define SD.

$$\mathrm{SD} = \frac{\sum |\mathrm{A} \oplus \mathrm{B}|}{|\mathrm{A}|} \qquad (16)$$

where A is a binary volume created by segmenting the SR volume and B is a binary volume created from the original volume. In order to create these segmented binary volumes, a thresholding technique is applied to the estimated super resolution volume and the original volume. This can be expressed as

$$I_T = \begin{cases} 1 & \text{if } I > T \\ 0 & \text{otherwise} \end{cases} \tag{17}$$

where T is the threshold value.

For a more standard evaluation, we compared the peak signal to noise ratio (PSNR) between the output volume and the original volume as a function of the iteration number. We use the following formula to define the PSNR.

$$\mathrm{PSNR} = 10\log_{10}\left(\frac{I_{\max}^2}{\mathrm{MSE}}\right) \tag{18}$$

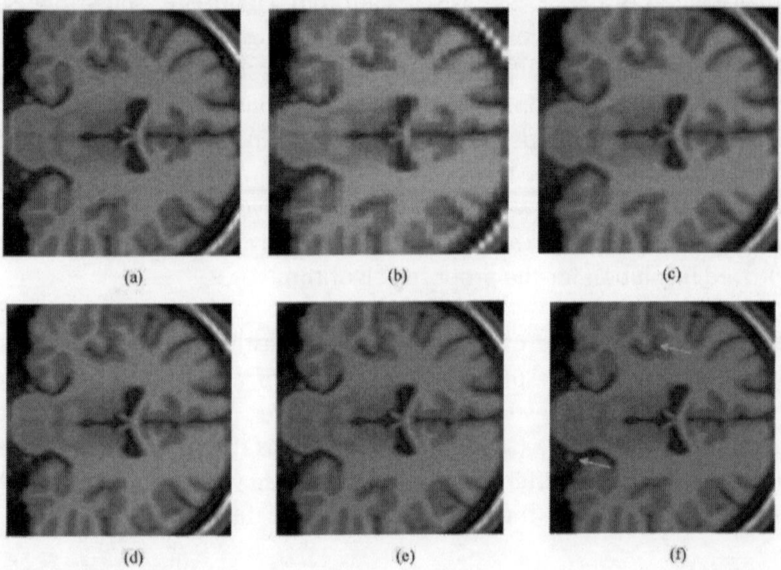

Figure 3. One slice from (a) The original volume, Original Slice; (b) One of the LR input volumes, Low resolution slice; (c) The output volume for the SR with DBTV algorithm, SR with DBTV; (d) The output volume from the SR with SART algorithm, SR with SART; (e) The output volume for the proposed SR with GSM constraint algorithm, Proposed GSM method; (f) The output volume for the proposed SR with sub-level GSM constraint algorithm, Proposed sub-level GSM method.

where I_{max} is the highest possible intensity value of the image volume and MSE is the mean squared error between the original volume and the output of the SR volume.

The results using these measurements for all algorithms are shown in **Figure 4** and **Table 1**.**Figure 4**(a) shows a plot of the SD as a percentage at each iteration when the white matter from the estimated SR volumes was segmented and compared to the segmented white matter from the original volume. **Figure 4**(b) shows the curves for PSNR of the estimated volume compared to the original volume for each SR technique. These results show that the performance of both our proposed SR algorithms which use a GSM constraint are superior to the alternative SR with DBTV and SR with SART algorithms and the sub-level GSM approach provides improved performance over the algorithm which uses the standard DTCWT. The PSNR values when compared with the original volume after the final iterations of each algorithm are shown in **Table 1**. The results in Figures 3 and 4 and **Table 1** clearly show the superior resolution improvement for this type of data when using our proposed methods.

5.2. Volume Reconstruction Using Real MRI Data

For a more practical evaluation, we tested our algorithm by using four sets of real knee MRI data of size 256 × 256 × 64 pixels. The slice thickness of each low-resolution volume was 2 mm and the in-plane resolution was 0.7 mm × 0.7 mm. The four data volumes used in our experiment were taken at different orientations. 3D rigid body registration was applied to align the femur in all four LR volumes. We then applied our proposed algorithm as well as the other algorithms described in the previous section to recover an SR volume from the set of LR volumes.

It is difficult to show quantitatively that SR has been achieved for real LR MRI data volumes. In this paper we chose to compare the shape of the segmented bones from the SR MRI volumes to the shape of the segmented bone from a CT volume of the same knee. Again we use the SD as defined in Section 5.1 as well as the dice similarity coefficient (DSC) as defined below:

$$DSC = \frac{2\sum|A \cap B|}{\sum|A| + \sum|A|} \tag{19}$$

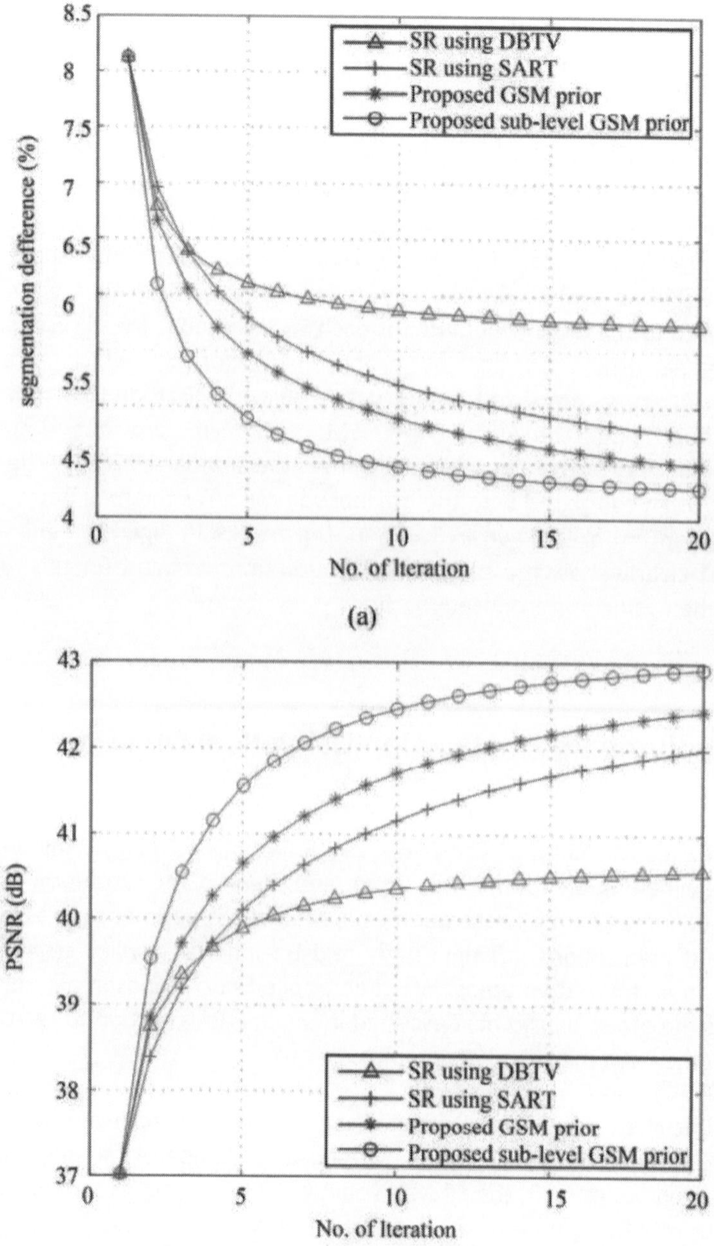

(b)

Figure 4. (a) The segmentation difference and (b) The PSNR for the SR algorithms using DBTV, SART and the proposed GSM and sub-level GSM model constraints.

Table 1. Final PSNR values for each algorithm (dB).

	Algorithms			
	DBTV	SART	GSM	Sub-level GSM
PSNR	40.5	41.8	42.4	42.9

The DSC determines the degree of overlap between the segmented output of the SR MRI volumes and the segmented CT volume. A higher value for the DSC measure indicates more similarity between two segmented volumes.

The slice thickness for the CT scan was 0.5 mm and the in-plane resolution was 0.335 mm × 0.335 mm. We then segmented the femur from the CT volume and from the SR MRI volume. It is difficult to segment the MRI femur using a fully automated segmentation algorithm because of noise and more than one tissue type in the MRI volume. Therefore the application of manual preprocessing steps were required to properly segment the output of the SR MRI volumes. The output slices of the SR MRI volumes were first manually delineated to select the boundary of the femur. An automated segmentation algorithm was then utilized to properly extract the femur from the semi-segmented SR MRI volumes. Finally a 3D multi-modal image registration method was applied to align the segmented bone from the CT volume and the segmented bone from the output of the SR MRI volumes. In our work, we used a 3D version of the inverse compositional algorithm [29] for image registration. The sum of conditional variance was used as the similarity measure instead of the sum of square difference as it has been shown to provide good results for multi-modal image registration [30]. Before the registration, we calibrated the CT image with the output of the SR MRI volume so that both image volumes were at the same scale. The SD and DSC measures were then calculated using the registered segmented femur from the SR MRI volumes and CT volume. The results for SD and DSC expressed as a percentage are shown in **Table 2**.

For a final analysis and evaluation, we measured the standard deviation of the difference between the surface of the segmented bone in the CT volume and the surface of the bone in the output of the SR MRI volumes.

The lower the standard deviation, the more closely the surface of the bone reconstructed from the MRI matches the surface obtained from the CT volume. The standard deviation of the surface distances between these two modalities is shown in **Table 3**.

Table 2. Final SD and DSC values for each Algorithm (%).

	Algorithms			
	DBTV	SART	GSM	Sub-level GSM
SD	19.1	17.0	13.5	12.9
DSC	88.7	89.8	90.4	91.3

Table 3. Standard deviation of the distance between the surfaces of the femur segmented from the CT and the LR and SR MRI volumes (mm).

	Algorithms				
	LR	DBTV	SART	GSM	Sub-level GSM
Standard deviation	1.39	1.32	1.32	1.28	1.26

Figure 5(a) shows a single-colour 3D rendering of the surface of the femur derived from the segmented CT volume. Figures 5 (b)-(f) show 3D renderings of the same CT surface but with colours used to illustrate the distance from the bone surface derived from the CT volume to the bone surfaces derived from the different MRI volumes. The colourbar in **Figure 5**(f) indicates the distance in millimeters that corresponds to each colour. Figures 5(b)-(f) show the distances between the surface derived from the CT volume and the surface derived from the LR input volume, the output of the SR algorithm using DBTV, the output of the SR algorithm using SART,

the output of our proposed SR algorithm using a GSM constraint and the output of our proposed SR algorithm using a sub-level wavelet transform and a GSM constraint respectively. Although there is some residual difference in the surfaces produced by the MRI and CT volumes, these surface renderings show the improvement provided by the proposed GSM model constraint and further improvement provided by using the sub-level wavelet transform.

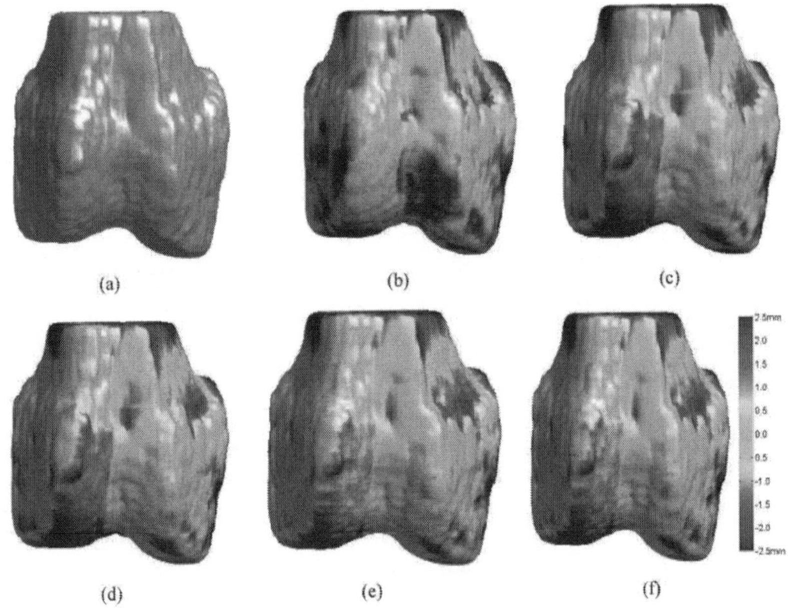

Figure 5. 3D surface renderings of the femur derived from (a) the segmented CT volume, CT Surface; (b) One of the LR MRI volumes, Low resolution volume; (c) The output volume for the SR with DBTV algorithm, SR with DBTV; (d) The output volume from the SR with SART algorithm, SR with SART; (e) The output volume for the proposed SR with GSM constraint algorithm, Proposed GSM method; (f) The output volume for the proposed SR with sub-level GSM constraint algorithm Proposed sub-level GSM method. The surface colours in (b)-(f) indicate the distance in millimeters between the surface derived from the CT volume in (a) and those obtained from the different MRI volumes.

6. CONCLUSION

In this paper, we have presented a novel SR algorithm where a sub-level wavelet transform is incorporated with the use of a GSM model wavelet sparseness constraint. The algorithm takes several multi-slice volumes of the same anatomical region captured at different orientations and combines these low-resolution volumes together to form a single 3D volume with much higher resolution in the slice-select direction. We have tested our algorithm using both synthetic and real MRI data both visually and quantitatively. From these error measurements, we can conclude that the proposed approaches provide a superior improvement in resolution when compared to the best performing previously proposed approaches.

REFERENCES

1. Huppert, B.J., Brandt, K.R., Ramin, K.D. and King, B.F. (1999) Single-shot fast spin-echo MR imaging of the Fetus: A pictorial essay. Radiographics, 19, 215-227. http: //radiographics.rsna.org/ content/ 19/suppl 1/S215

2. Vignaux, O.B., et al. (2001) Comparison of single-shot fast spin-echo and conventional spin-echo sequences for MR imaging of the heart: Initial experience. Radiology, 219, 545-550. http://radiology.rsna. org/ content/219/2/545

3. Gholipour, A., Estroff, J. and Warfield, S. (2010) Robust super-resolution volume reconstruction from slice acquisitions: Application to fetal brain MRI. IEEE Transactions on Medical Imaging, 29, 1739-1758.

4. Greenspan, H. (2009) Super-resolution in medical imaging. The Computer Journal, 52, 43-63

5. Plenge, E., et al. (2012) Super-resolution methods in MRI: Can they improve the trade-off between resolution, signal-to-noise ratio, and acquisition time? Magnetic Resonance in Medicine, 68, 1983-1993.

6. Park, S.C., Park, M.K. and Kang M.G. (2003) Superresolution image reconstruction: A technical overview. IEEE Signal Processing Magazine, 20, 21-36.

7. Peled, S. and Yeshurun, Y. (2001) Super resolution in MRI: Application to human white matter fiber tract visualization by diffusion tensor imaging. Magnetic Resonance in Medicine, 45, 29-35.

8. Greenspan, H., Peled, S., Oz, G. and Kiryati N. (2001) MRI inter-slice reconstruction using super-resolution. Prceedings of Medical Image Computing and ComputerAssisted Intervention MICCAI 2001, Lecture Notes in Computer Science, 2208, 1204-1206.

9. Irani, M. and Peleg, S. (1991) Improving resolution by image registration. CVGIP: Graphical Models and Image Processing, 53, 231-239.

10. Scheffler, K. (2002) Superresolution in MRI? Magnetic Resonance in Medicine, 48, 408-408.

11. Kramer, D., et al. (1990) Applications of voxel shifting in magnetic resonance imaging. Investigative Radiology, 25, 1305-1310.

12. Greenspan, H., Oz, G., Kiryati, N. and Peled, S. (2002) Super-resolution in MRI. Proceedings of the IEEE International Symposium on Biomedical Imaging, Washington DC, 7-10 July 2002, 943-946.

13. Yan, Z. and Lu, Y. (2009) Super resolution of MRI using improved IBP. Proceedings International Conference on Computational Intelligence and Security, Beijing, 11-14 December, 643-647.

14. Peeters, R.R., et al. (2004) The use of super-resolution techniques to reduce slice thickness in functional MRI. International Journal of Imaging Systems and Technology, 14, 131-138.

15. Joshi, S., et al. (2009) MRI resolution enhancement using total variation regularization. Proceedings IEEE International Symposium on Biomedical Imaging: From Nano to Macro, Boston, 28 June - 1 July 2009, 161-164.

16. Ben-Ezra, A., Greenspan, H. and Rubner Y. (2009) Regularized super-resolution of brain MRI. Proceedings IEEE International Symposium on Biomedical Imaging: From Nano to Macro, Boston, 28 June - 1 July 2009, 254-257.

17. Shilling, R., et al. (2009) A super-resolution framework for 3-D high-resolution and high-contrast imaging using 2-D multi-slice MRI. IEEE Transactions on Medical Imaging, 28, 633-644.

18. Islam, R., Lambert, A.J. and Pickering, M.R. (2012) Super resolution of 3D MRI images using a Gaussian scale mixture model constraint. Proceedings IEEE International Conference on Acoustics, Speech and Signal Processing, Kyoto, 25-30 March 2012, 849-852.

19. Kingsbury, N. (2001) Complex wavelets for shift invariant analysis and filtering of signals. Applied and Computational Harmonic Analysis, 10, 234-253.

20. Bertero, M. and Boccacci, P. (1998) Introduction to inverse problems in imaging. Taylor & Francis, Bristol.

21. Landweber, L. (1951) An iteration formula for fredholm integral equations of the first kind. American Journal of Mathematics, 73, 615-624.

22. Daubechies, I., Defrise, M. and De Mol, C. (2004) An iterative thresholding algorithm for linear inverse problems with a sparsity constraint. Communications on Pure and Applied Mathematics, 57, 1413-1457.

23. Vonesch, C. and Unser, M. (2008) A fast thresholded landweber algorithm for wavelet-regularized multidimensional de-convolution. IEEE Transactions on Image Processing, 17, 539-549.

24. Zhang, Y. and Kingsbury, N. (2009) Image deconvolution using a Gaussian Scale Mixtures model to approximate the wavelet sparseness constraint. Proceedings of IEEE International Conference on Acoustics, Speech and Signal Processing, Taipei, 19-24 April 2009, 681-684.

25. Sendur, L. and Selesnick, I. (2002) Bivariate shrinkage functions for wavelet-based denoising exploiting interscale dependency. IEEE Transactions on Signal Processing, 50, 2744-2756.

26. Andrews, D.F. and Mallows, C.L. (1974) Scale mixtures of normal distributions. Journal of the Royal Statistical Society. Series B (Methodological), 36, 99-102. http://www.jstor.org/stable/2984774

27. Wainwright, M.J. and Simoncelli, E.P. (2000) Scale mixtures of Gaussians and the statistics of natural images. Advanced Neural Information Processing Systems, 12, 855-861.

28. Collins, D., et al. (1998) Design and construction of a realistic digital brain phantom. IEEE Transactions on Medical Imaging, 17, 463-468.

29. Baker, S. and Matthews, I. (2004) Lucas-Kanade 20 years on: A unifying framework. International Journal of Computer Vision, 56, 221-255.

30. Pickering, M.R. (2011) A new similarity measure for multi-modal image registration. Proceedings of IEEE International Conference on Image Processing, Brussels, 11-14 September, 2011, 2273-2276.

Chapter 6

On Wavelet Transform General Modulus Maxima Metric for Singularity Classification in Mammograms

Tomislav Bujanovic[1], Ikhlas Abdel-Qader[2]

[1]Department of Electrical Engineering & Computer Science, Syracuse University, Syracuse, USA

[2]Department of Electrical & Computer Engineering, Western Michigan University, Kalamazoo, USA

ABSTRACT

Continuous wavelet transform is employed to detect singularities in 2-D signals by tracking modulus maxima along maxima lines and particularly applied to microcalcification detection in mammograms. The microcalcifications are modeled as smoothed positive impulse functions. Other target property detection can be performed by adjusting its mathematical model. In this application, the general modulus maximum and its scale of each singular point are detected and statistically analyzed locally in its neighborhood. The diagnosed microcalcification cluster results are compared with health tissue results, showing that general modulus maxima can serve as a suspicious spot detection tool with the detection performance no significantly sensitive to the breast tissue background properties. Performed fractal analysis of selected singularities supports the statistical findings. It is important to select the suitable computation parametersthresholds of magnitude, argument and frequency range-in accordance to mathematical description of the target property as well as spatial and numerical resolution of the analyzed signal. The tests are performed on a set of images with empirically selected parameters for 200 μm/pixel spatial and 8 bits/pixel numerical resolution, appropriate for detection of the suspicious spots in a mammogram. The results show that the magnitude of a singularity general maximum can play a significant role in the detection of microcalcification, while zooming into a cluster in image finer spatial resolution both magnitude of general maximum and the spatial distribution of the selected set of singularities may lead to the breast abnormality characterization.

KEYWORDS

Continuous Wavelet Transform; Fractal Dimension; General Modulus Maximum; Microcalcification; Singularity; Smoothed Impulse Function

1. INTRODUCTION

Wavelet transform modulus maxima method was developed for detection and characterization of signal singularities by Mallat and his collaborators [1-3]. Their method detects signal singularities by tracking the wavelet coefficients magnitude maximum over scale. They proved that, if a wavelet function is derivative of a Gaussian, wavelet transform modulus maxima must propagate towards finer scales. Although the representation by discrete wavelet maxima is not complete since several signals may exhibit the same wavelet maxima [4], Mallat's numerical experiments have shown that it is possible to reconstruct signals with a relatively small mean square error (smaller than 1%) [3].

Arneodo and his team focused on how to recognize a sharp signal transition by tracking its behavior over scale. Specifically algorithms based on continuous wavelet transform modulus maxima method are able to detect singular points in a discrete 2-D and 3-D signals and supported by fractal analysis to give the metrics for the local signal regularity [5-9]. Arneodo's team developed fractal based algorithm [8] supported by modulus maxima method to analyze turbulent 2-D and 3-D signals.

Microcalcifications in breast are residual calcium deposits that originate not only from completely normal processes, but also abnormal ones. Shape, morphology, and spatial distribution of individual microcalcifications are some of the features detectable in X-ray mammograms that suggest benign or malignant breast abnormality. Researchers have made few breast tissue classifications and have recognized more than twenty of those conventional features [10-12]. One acceptable simplification is to describe microcalcifications as ellipsoids of diameters between 0.05 mm and 1 mm. For early cancer detection, calcifications with spatial extent less than 0.5 mm are most important for clinical diagnosis. Particularly, this corresponds to calcifications roughly in the order of 0.1 mm in diameter. Also, microcalcifications appearing in clusters may suggest malignancy while individual occurrences are of low clinical significance [13].

Wavelets as mathematical microscope have been used by many researchers in mammogram analysis. Wang and Karayiannis [14] proposed searching for microcalcifications by using high frequency energy blobs in wavelet decomposed mammograms. But normal signal fluctuation in mammograms may have dominant energy blobs in some scales making a mask over significant abnormality related information. Indeed, an energy blob is just a location where further analysis should follow. We use modulus maxima method to eliminate the singularities with positive Hölder exponent, because those singularities are present everywhere in mammograms (this is commented by fractal dimension of singularity spatial distribution in Section 4.4). In this work we show that the general modulus maximum, the largest magnitude value along a maxima line, and the scale in which it is detected are the most significant information to be used in further detection procedure. The argument of the detected general modulus maxima (the tangential directions of an edge around pixel) is another significant piece of information that is very useful in the target segmentation procedure (not discussed in this work).

Zhang and coworkers [15] studied the size of microcalcifications and developed an optimally weighted wavelet transform method by which, before image reconstruction, they multiplied dyadic wavelet coefficients by optimized coefficients 0.4, 1.6, 1.0, and 0.05 for the scales corresponding to 100 μm, 200 μm, 400 μm, and 800 μm respectively. They achieved $A_z = 0.92$ under ROC curve, outperforming $A_z = 0.86$ for difference-image technique.

Microcalcifications in mammogram analysis may be modeled as impulse functions smoothed by Gaussian filters. Strickland and Hahn [16] recognized Gaussian nature of microcalcifications spatial intensity and applied 2-D Gaussian filters for microcalcification detection. They inserted inter-scales to increase detection sensitivity for wide range of possible microcalcifications.

Since the Food and Drug Administration approved the technology in 1998, several Computer-Aided Detection (CAD) systems that support microcalcification detection have been deployed for clinical use. Literature reports show mixed results of current CAD systems in practice [17,18]. Results of some of the studies performed by using retrospective analyses or computer modeling suggested that CAD can achieve the main task increasing cancer detection.

Some studies show no practical improvements concluding that the CAD effect on the accuracy of interpretation is unclear [19] mostly because of radiologist's false sense of confidence in CAD although in testing a CAD system applied to mammograms with known cancers might have excellent performance, known as "laboratory effect" [20].

This leads to a conclusion that different conditions of mammographic characteristics associated with X-ray exposure and breast tissue density should be studied across different institutions [21].

In this work, we consider microcalcifications as sharp signal transitions modeled as impulse functions smoothed by a Gaussian and characterized by negative Hölder exponent. We detect smoothed impulse functions and identify general modulus maxima and the scales in which they appear. Essentially detected general maxima show the scale in which the singularities are visible as the step functions with Hölder exponent approaching to zero. If in the higher scales the magnitude of the wavelet coefficients starts decreasing the singularity shows its impulse nature. **Figure 1** illustrates the modulus maxima behavior over scale for the singularities with and without impulse nature, i.e. microcalcification and normal breast tissue signal fluctuation. The magnitude of modulus maxima of normal breast tissue signal fluctuation can become dominant (higher energy in higher scales) and affect the detection of the singularity with impulse nature (lower energy in higher scales) which represents a microcalcification in this case.

We detect the wavelet coefficient magnitude in the scale in which a singularity behaves as a step function and in higher scales becomes negative showing impulse nature of the singularity. Notice that normal breast tissue exhibits local signal fluctuation with many smoothed impulse functions in mammograms. This requires that in the next step detected smoothed impulse functions should be classified as significant for further analysis if they have locally dominant magnitudes of general maxima. In short, we take advantage of the capabilities of continuous wavelet transform modulus maxima method to detect signal singularities and differentiate between microcalcifications and normal signal fluctuation keeping track how Hölder exponent is developing over scale.

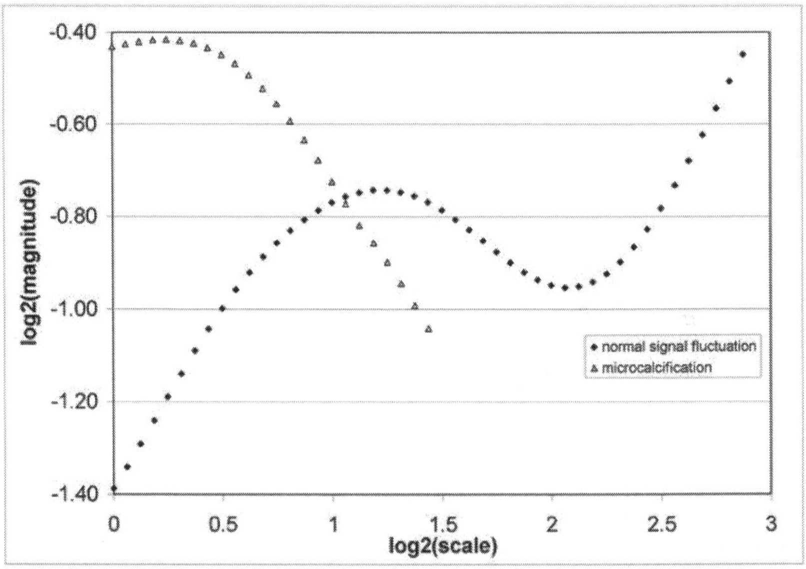

Figure 1. Typical log-log characteristics Modulus Maxima vs scale for a normal signal fluctuation (¨) and a microcalcification (Δ) in the mammogram mdb226 [22].

After detection in coarser spatial resolution, based on the fact that this method can detect step function, a cluster of singularities could be segmented in finer resolutions with appropriately adjusted settings. This method could be used for singularity border transition analysis i.e. its shape analysis for diagnostic purposes and therefore microcalcification estimation malignant vs benign [12].

In this work we demonstrate use of general maxima metric in mammograms with 200 μm/pixel spatial and 8 bits/pixel numerical resolution to speed up detection not of microcalcification border, but microcalcification itself. The scale of the general maxima detection depends on the size of the singularity and since singularity size in mammograms differs from one to another there is no typical scale but scale distribution that could be identified.

The magnitudes and their spatial distribution are analyzed in 25 regions containing microcalcification clusters from diagnosed 20 mammograms and the results compared to the results of 20 regions with healthy tissue from the same mammograms [22]. Additionally, we recognize and select smoothed impulse function by its negative Hölder exponent. Detection of a magnitude of a general maximum doesn't depend on the negative slope value within the detection procedure. This makes general modulus maximum stronger and more independent feature i.e. that is an excellent candidate to integrate with other features.

Background on the continuous wavelet transform, modulus maxima method, and fractal dimensions are presented in Section 2. In Section 3 we present the details of the framework and its implementation. The results from both statistical and fractal implementations, including the samples of tested mammograms, are presented in Section 4 while Conclusions are presented in Section 5.

2. BACKGROUND

The presence of strong local singularities is characterized by negative Hölder exponent while normal fluctuation is mostly characterized by positive Hölder exponent. This makes an isolated singularity detectable by analyzing the behavior of a modulus maximum along its maxima line.

Mallat and Hwang [1] proved that the magnitudes of the wavelet coefficients are bounded as shown in (1):

$$\left|W_a f(x,y)\right| \leq A \cdot a^{\alpha} \tag{1}$$

where $|W_a f(x,y)|$ is magnitude of the wavelet coefficient at point (x, y) in scale a, a is Hölder exponent at (x, y) and A > 0 is a constant.

When Hölder exponent at (x, y) is positive the magnitude of wavelet coefficient will increase with scale while when it is negative, as in case of impulse function, the magnitude of the wavelet coefficient will decrease. Mathematically local microcalcification of the mammogram signal may be modeled as an impulse function convolved by a Gaussian smoothing function.

For further analysis of behavior of magnitude of wavelet coefficients in scale, we use a wavelet coefficient magnitude approximation [3]

$$\log_2|Wf(u,a)| \approx \log_2(A)+\left(a+\frac{1}{2}\right)\log_2 a-\frac{n-a}{2}\log_2\left(1+\frac{\sigma^2}{\beta^2 a^2}\right) \tag{2}$$

where a targeted microcalcification is modeled as an impulse function smoothed by Gaussian of variance σ^2 and convolved by a Gaussian wavelet with variance β^2.

If a tends to −1, then the magnitude of a wavelet coefficient increases up to the scale corresponding to the variance s^2 reaching wavelet transform general modulus maximum followed by a decrease as in the case of microcalcification shown in **Figure 1**. An example of normal signal fluctuation generally characterized by positive Hölder exponent in the scales having increasing magnitude of wavelet coefficients is also shown in **Figure 1**.

2.1. Modulus Maxima Representation

A Gaussian smoothing function at scale a

$$\phi_a(x,y)=\frac{1}{2\pi(a\sigma_0)^2}\exp\left(-\frac{x^2+y^2}{2(a\sigma_0)^2}\right) \tag{3}$$

has been used as a kernel for wavelet transform discrete filters. The horizontal and vertical wavelet coefficients at scale a are defined as

$$W_a^{hor}(f) = f * \Psi_a^{hor} \tag{4}$$

And

$$W_a^{ver}(f) = f * \Psi_a^{ver} \tag{5}$$

respectively, where the horizontal and vertical wavelets Ψ_a^{hor} and Ψ_a^{ver} in scale a are derivatives of smoothing function f_a in the same scale a

$$\Psi_a^{hor}(x,y) = \frac{\partial \phi_a(x,y)}{\partial x} \tag{6}$$

And

$$\Psi_a^{hor}(x,y) = \frac{\partial \phi_a(x,y)}{\partial y} \tag{7}$$

respectively.

Wavelet coefficient at a scale a is represented by its magnitude and argument as follows,

$$W_a(f) = (M_a(f), A_a(f)) \tag{8}$$

where $M_a(f) = \sqrt{(W_a^{hor}(f))^2 + (W_a^{ver}(f))^2}$ and $A_a(f) = Arg(W_a^{hor}(f) + j \cdot W_a^{ver}(f))$, with horizontal and vertical wavelet coefficients W_a^{hor} and W_a^{ver} at scale a, defined in (4) and (5) respectively.

A modulus maximum at (x_0,y_0) at the scale a is represented by its magnitude if it is equal or greater than wavelet coefficient magnitude in its neighborhood. There is no detected modulus maximum at (x_0,y_0) in scale a if any surrounding wavelet coefficient magnitude is grater than magnitude at (x_0,y_0), i.e.

$$MM_a(x_0,y_0) = \begin{cases} M_a & M_a(x_0,y_0) \geq \max\limits_{(x,y)\in\aleph_a(x_0,y_0)} M_a(x,y) \\ 0 & \text{otherwise} \end{cases} \quad (9)$$

where $\aleph_a(x_0,y_0)$ is 8-neighborhood of (x_0,y_0) in the scale a.

While a maxima line consists of the wavelet transform modulus maxima corresponding to the same local feature but in different scales, the general modulus maximum is a point on the maxima line characterized by its maximal magnitude $gM(x_0,y_0)$ and the scale $a_{gM}(x_0,y_0)$ in which it is detected. Wavelet transform general modulus maximum will appear with a magnitude and in a scale that depends on the initial variance of the smoothing Gaussian in the initial scale $a_0 = 1$.

In the higher scales of $a_{gM}(x_0,y_0)$ corresponding maxima line will exist as long as the smoothed impulse function is the dominant singularity in its neighborhood. In that range the slope of the log-log diagram modulus maxima vs scale will tend to −1 if the maxima line is long enough, i.e. until it is overlapped by a modulus maximum of another dominant singularity. If in higher scales the singularity stops being dominant in its neighborhood, the maxima line tracking process should be interrupted to prevent misleading singularity detection. The magnitude $gM(x_0,y_0)$ and the scale $a_{gM}(x_0,y_0)$ in which the general modulus maximum is detected as well as the length of maxima line afterward are significant starting points for evaluating the isolated singularities.

2.2. Fractals, Partition Function and Dimension

A fractal is an object or quantity that displays self-similarity in all scales. A plot of the quantity on a log-log graph versus scale then gives a straight line, which slope is said to be the fractal dimension. The prototypical example for a fractal is the length of a coastline measured with different length ruler. The shorter the ruler, the larger the length measured, a paradox known as the coastline paradox or “the Richardson effect” (L. F. Richardson, 1881-1953) [23].

Fractal dimension or capacity dimension of a fractal is the exponent D in $n(\varepsilon) \propto \varepsilon^{-D}$, where $n(\varepsilon)$ is the minimal number of balls of diameter e needed to cover the fractal set in each scale. Wavelets provide a natural generalization of the classical box-counting techniques to fractal signals, i.e. the wavelets are playing the role of generalized boxes. The wavelet transform as a mathematical microscope can be used to extract microscopic information about scaling properties of fractal objects [6]. Partition function can be derived from maxima line skeleton as

$$Z(q,a) = \sum \left(\sup_{L,a' \leq a} M_{\Psi}\left[f\left((x,y),a'\right) \right] \right)^{q}, q \in \Re \tag{10}$$

The analogy related to multifractal formalism in thermodynamics allows the following definition of the exponent t(q) for the behavior of partition function along the scale:

$$Z(q,a) \propto \alpha^{\tau(q)}, a \to 0^{+} \tag{11}$$

where q and t(q) represent inverse temperature and free energy in statistical mechanics. For homogenous fractal functions-monofractals and q = 0, the value of –t(q) is associated to fractal dimension of the set of all selected singularities [7].

According to (11), fractal dimension of a selected set of singularities is determined as the negative value of the slope of log-log characteristics of partition function Z(q,a) vs scale a with q = 0 as shown in Section 4.

3. FRAMEWORK

In our experiments the targets are the microcalcifications sized from 0.2 mm to 1 mm in diameters. Accordingly to the mammogram spatial resolution, the scale is changing in three octaves. Notice that algorithm parameters, such as initial smoothing function variance and length of maxima lines necessary for the target detection, depend on the mammogram spatial resolution and the size of expected targets. Also, the general modulus maxima analysis can be used in 1-D and 3-D signals in a similar fashion.

Changing the scale, Gaussian filters are used to compute wavelet transform coefficients of mammographic images in vertical and horizontal direction in three octaves. The framework is based on using wavelet transform modulus maxima to identify the maxima lines in a mammogram and then extract the general modulus maximum on each maxima line. Also in each maxima line a corresponding minimum is detected in higher scales than general maximum and before maxima line fades out. Singularities pointed by maxima lines with no identified corresponding minimums are considered as normal signal fluctuation and rejected from further analysis. Singularities with detected corresponding minimums have negative Hölder exponents. The fact that described general maximum and corresponding minimum are clearly detected means the log-log characteristic modulus maxima vs scale has negative slope and isolated smoothed impulse function is localized at the point where maxima line propagates to scale $a = 1$. A maxima line might not propagate to a single point at scale $a = 1$ because of fast oscillations in the cone of influence around terminal point of the maxima line in the lowest detected scale [3]. This location will be classified as an eligible singularity if the general maximum and corresponding minimum is detected.

An illustration of applied framework is shown in Figure 2.

Can we fix that now? Should we reformat the entire text to avoid empty (blank) space or they would do it?

Variance of smoothed impulse function can be estimated by the scale in which general maximum is detected [24]. The selection of wavelet coefficients depends on the description of the singular point that we are detecting. In mammogram signals, we are adjusting the scale to the expected size of the microcalcification (50 μm to 1 mm) and the input spatial resolution of the mammogram (usually 50 μm). If both the general maximum and corresponding don't exist on a maxima line its detected singularity is classified as normal signal fluctuation, i.e. the maxima line is not related to a microcalcification. A microcalcification boundary can be detected in the scales that correspond to 50 mm to 100 mm, i.e. the microcalcification segmentation needs additional information from finer image spatial resolution and it was not part of these experiments. Notice that, for example, the shape [12] of the targeted singularity can be involved in further singularity sub-classification.

Gaussian kernels and their derivatives are used to enable image processing with no particular bias in the spacescale analysis. Significant fractal properties can change very fast in scale. They often become invisible within less than a 1/2 of an octave and with coarse scale increment the trend of modulus maxima behavior in scale could not be detected.

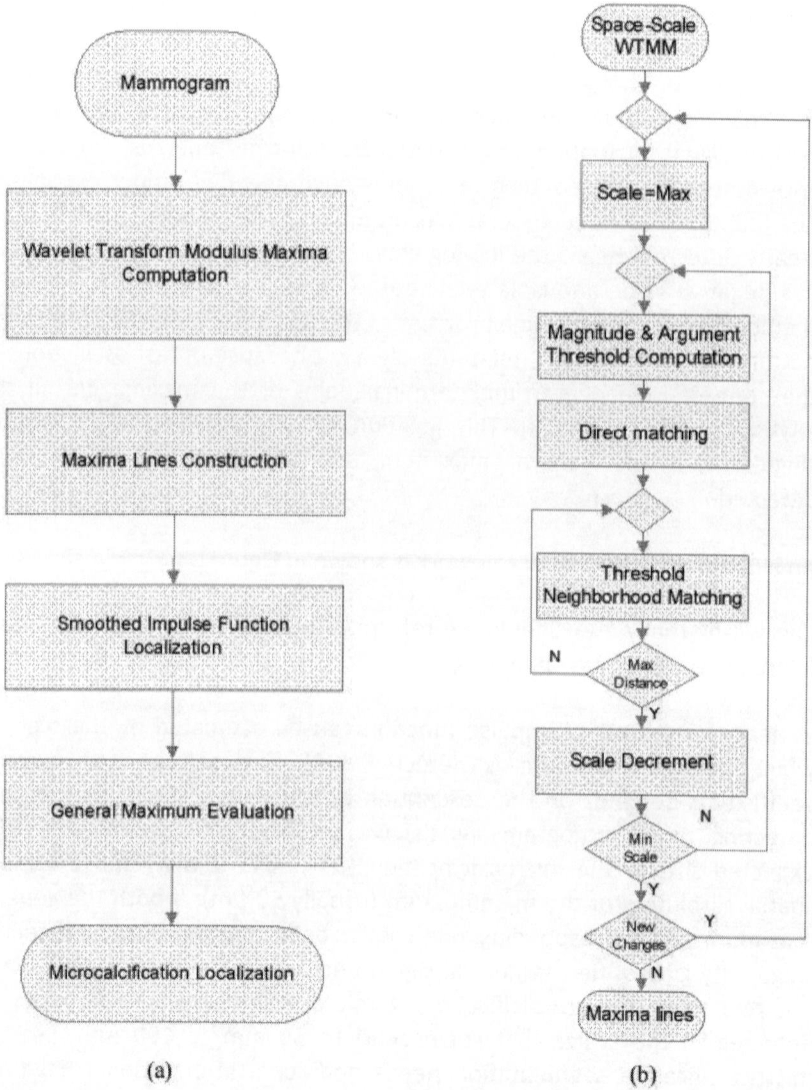

Figure 2. General Maximum detection and microcalcification localization block diagram (a) and maxima line construction block diagram (b).

Each maxima line points to a singularity and each singularity is analyzed based on the modulus maxima behavior along its maxima line.

Detected singularities gradually change in scale both the magnitude M and the argument q of their modulus maxima. Two maxima points in two consecutive scale layers belong to same maxima line if their spatial locations and their arguments are similar. The level of similarity of spatial distance and argument difference are below empirically determined thresholds. The experiments show that the detection process converges if the scale increment between two consecutive scale layers is as low as inc = $2^{1/16}$. Then the algorithm will properly form a maxima line with the argument threshold dA = p × $\log_2$(inc)/4 radians or Dq < dA = p/64 radians in the simulations presented in Section 4.

In some specific situations when maxima line switches from one point to another and both are the parts of the same feature, the argument difference can be significantly larger than the argument threshold and then magnitude threshold will be checked as an alternative to the argument threshold. The magnitude threshold applied in the simulations presented in this work was adaptive in both space and frequency domain, i.e. the threshold was $dM_{a0} = 3 \times abs(M_{a1} - M_{a2})$, where M_{a1} and M_{a2} are the modulus maxima detected in two higher consecutive scales. An example of described situation is presented in **Figure 3**.

With the parameter setup applied in this work we consider 1/4 of an octave a minimum scale distance between general modulus maximum and corresponding minimum to classify a location as an eligible singularity.

4. RESULTS AND DISCUSSION

The mammograms from database [22] are digitized at 50 μm pixel edge and reduced to spatial resolution 200 μm pixel edge and clipped/padded so that every image is 1024 × 1024 pixels. The area of a pixel represents 0.04 mm^2 of breast tissue. Numerical resolution is 8 bits/pixel. In this work 12 malignant and 13 benign microcalcification clusters detected in 20 mammograms along with other 20 healthy regions selected from the same mammograms are used.

We find that normal and noise related detected eligible singularities are represented with significantly lower general modulus maxima comparing to those related to diagnosed microcalcification. We have introduced spatially local adaptive threshold in magnitude, to suppress false positive findings. The magnitudes above the local threshold we name (locally) dominant magnitude.

In this work we study magnitude and spatial distribution of general maxima of three different sets of detected singularities in the 64 × 64 pixel sub-images of the diagnosed 25 microcalcification clusters:

1) All detected singularities which show negative Hölder exponent resembling smoothed impulse function

2) The subsets of the set selected in 1, with dominant magnitudes of computed general maxima resembling microcalcifications, and

3) The sets of singularities that correspond to the diagnosed microcalcifications in 25 clusters.

Assuming that the nature of malignant and benign microcalcification clusters is different we additionally study each category described above separately as malignant and benign abnormality.

In addition we study magnitude and spatial distribution of general maxima of three different sets of detected singularities in 20 healthy areas of same mammograms:

1) All detected singularities which show negative Hölder exponent and resembling smoothed impulse function in healthy areas, and

2) The subsets of 1 that have dominant magnitudes of computed general maxima in health areas.

Finally, we compare our results with respect to the background breast tissue type: dense (D), glandular (G), and fatty (F) tissue.

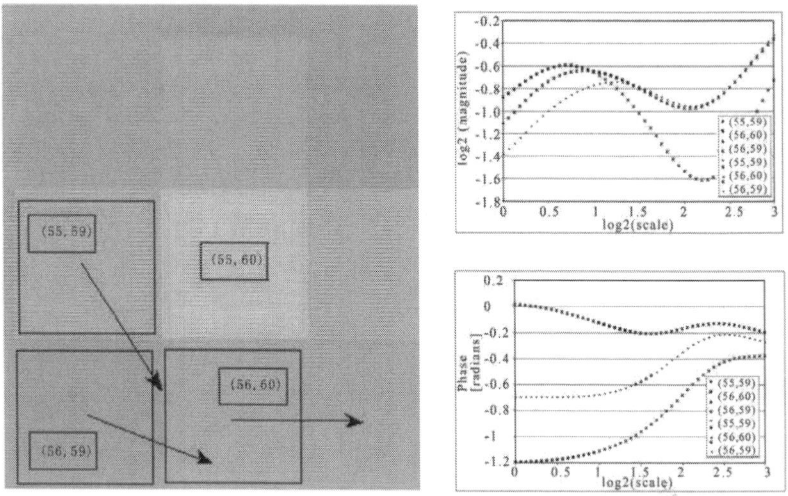

(a) (b)

Figure 3. Example of maxima line construction from three singular points that belong to same sharp transition: Three singular points around the pixel (55, 60), (a), that can compose a maxima line as shown in bold and colored symbols in (b-up). Their arguments are significantly different representing the tangential directions of an edge around pixel (55, 60) but in higher scales they converge to the same limit as clearly illustrated in (b-down). Bolder symbols in the diagrams represent the values that belong to detected maxima line representing the same higher scale structure.

4.1. Differentiation Normal vs Abnormal Singularities

In 25 microcalcification clusters our method detected total of 3115 eligible maxima lines. The histogram of eligible general maxima magnitudes is shown in **Figure 4**- blue line. The histogram of general maxima gM related to the diagnosed microcalcification findings is red line in **Figure 4**.

For the comparison, the histograms of the detection of eligible general maxima magnitudes in 20 healthy areas is determined and presented as a dashed black line in **Figure 4**, while the black line in the same diagram is the histogram of detected eligible general modulus maxima magnitudes with associated local intensity maximums in the neighborhood.

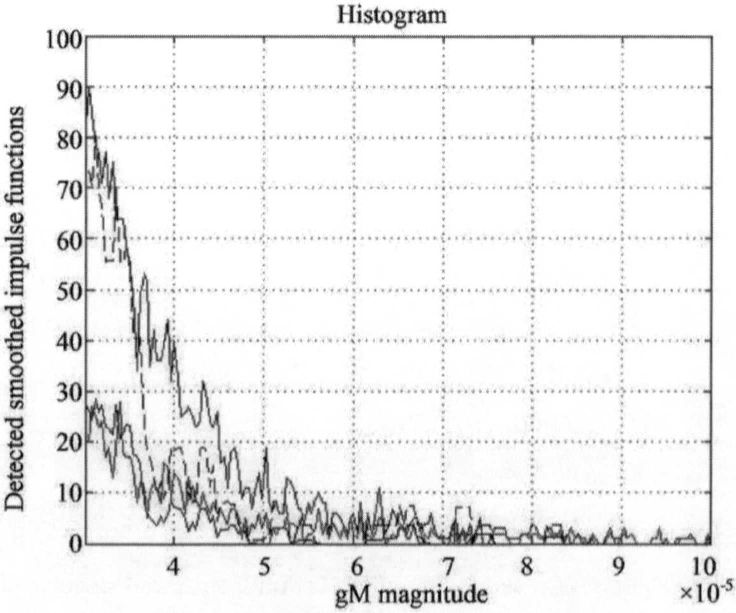

Figure 4. Histogram of total eligible findings in the areas of diagnosed clusters-blue line, and diagnosed microcalcifications in 25 clusters-red line. Histogram of total singularity findings in 20 healthy areas-black dashed line, and the eligible findings with a local intensity maximum in their neighborhood in same 20 healthy areas-black line.

For the purpose of further analysis of the diagrams in **Figure 4** we introduce diagnosed rate function $DRF_{\text{diagn}}(x)$ for the diagnosed microcalcifications by

$$DRF_{\text{diagn}}(x) = \frac{CDF_{\text{diagn}}(x)}{CDF_{\text{total}}(x)} \tag{12}$$

where x is general maxima magnitude,

$CDF_{\text{diagn}}(x) = \sum_{0}^{x} hist_{\text{diagn}}(x)$ is cumulative distribution function of general maxima for diagnosed microcalcifications (red line) and $CDF_{\text{total}}(x) = \sum_{0}^{x} hist_{\text{total}}(x)$

is cumulative distribution function of general maxima for total eligible general modulus maxima (blue line) both in [0, x].

While an impulse may be negative, a microcalcification is, in general, a spot brighter than its surroundings. Similarly to the function $DRF_{diagn}(x)$, here we institute the function $DRF_{local\_max}(x)$, the ratio of detected impulses with the local intensity maximums in their neighborhood and total detected smoothed impulse functions. i.e.

$$DRF_{local\_max}(x) = \frac{CDF_{local\_max}(x)}{CDF_{total}(x)} \tag{13}$$

where x is general maxima, magnitude,

$CDF_{local\_max}(x) = \sum_{0}^{x} hist_{local\_max}(x)$ is cumulative distribution function of general maxima for smoothed impulses with detected local maximum in each of their neighborhood (red line) and $CDF_{total}(x) = \sum_{0}^{x} \mathrm{hist}_{total}(x)$ is cumulative distribution function of general maxima for total eligible general modulus maxima (blue line).

In the same set of 20 mammograms we also choose 20 healthy areas and select all singularities that could be modeled as smoothed impulse function. The histogram of the total number of selected singularities vs gM magnitude is shown as dashed black line in **Figure 4**. Assuming that a microcalcification is presented in mammograms as a spot brighter than its neighborhood we select a sub-set of all singularities in which neighborhood a local intensity maximum is detected. The histogram of the singularities in the sub-set vs gM magnitude is shown as black line in **Figure 4.**

The diagram of the function $DRL_{local\_max}(x)$ in the set of analyzed sub-images with diagnosed microcalcification clusters is the black line in **Figure 5**, while the dashed black line in **Figure 5**represents same function $DRL_{local\_max}(x)$ but in the set of analyzed sub-images without diagnosed microcalcifications.

For each line in **Figure 5** we notice two segments: an increasing, say non-linear, part on the left-hand side and an approximately linear behavior of each line on the right hand side. Numerically, we notice that linearity of dashed black line (having negative slope) starts between 3×10^{-5} and 4×10^{-5} in the case of healthy tissue samples, while in the case of diagnosed abnormalities for both lines the linearity (red and full black lines (having positive slope) starts above 4×10^{-5}.

To confirm this observation about positive and negative slope we performed same experiment separating the results for dense (D), glandular (G), and fatty (F) tissue. In **Figure 6** the diagnosed rate functions $DRL_{diagn}(x)$ are presented as full colored lines: blue-glandular, reddense, and green-fatty background tissue. All three background tissue based full lines have increasing linear parts on the right hand side. The detection rate functions $DRF_{local\_max}(x)$ of smoothed impulse functions with local intensity maximums in the neighborhood is shown as dashed lines with same color schedule. All three dashed lines based on healthy tissue have decreasing linearity on the right hand side.

In same experiment we notice the difference in behavior of the function $DRF_{local\_max}(x)$ in the normal subimages (no microcalcifications-**Figure 7**(a) and in the subimage with diagnosed microcalcification clusters-**Figure 7**(b).

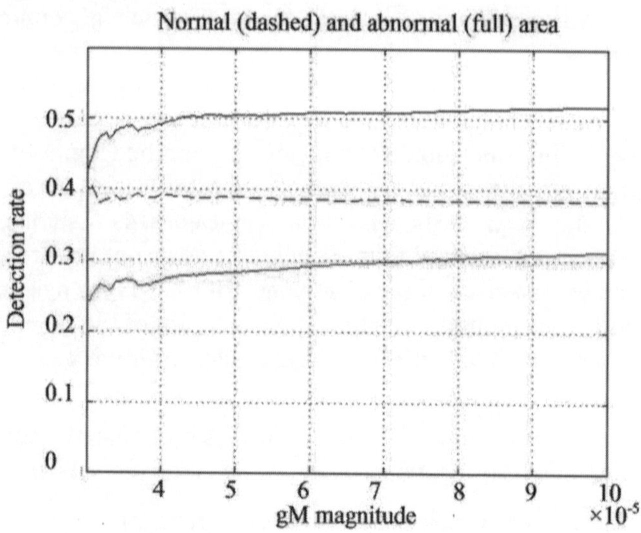

Figure 5. Diagnosed rate function $DRF_{diagn}(x)$ red line, detection rate of eligible singularities assiciated with a local intensity maximum in the neghborhood in the clusters $DRF_{local\_max}(x)$ full black line, and $DRF_{local\_max}(x)$ in healthy tissue dashed black line.

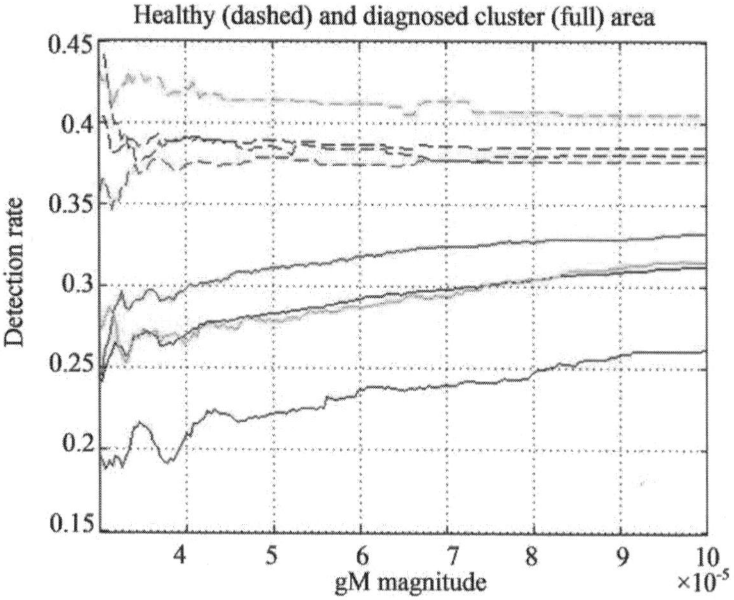

Figure 6. Smoothed impulse function detection rate comparison for diagnozed microcalcification clusters $DRF_{diagn}(x)$ (full lines) and for corresponding healthy tissue $DRF_{local\_max}(x)$ (dashed lines). Black lines show the behavior of detection rate vs general maxima magnitude for all tested clusters, while blue, red, and green lines are related to glandular, dense, and fatty tissue, respectively.

We conclude that positive vs. negative slope of the linear parts on the right hand side leads to a good differentiation between abnormal and normal tissue sample. We notice that the average magnitude of the general modulus maxima in 25 clusters is 4.18×10^{-5}, while in 20 healthy background sample is 3.59×10^{-5}. Both averages are in the range of gM magnitude that could be considered in either non-linear or linear part of corresponding lines. This leads to another conclusion that a successful differentiation normal vs. abnormal findings could be expected when an adaptive threshold is set to the local average of gM magnitudes.

For the abnormalities below the local averages the additional microcalcification features should be considered.

4.2. Threshold

In tested 25 clusters of microcalcifications we found 3115 eligible maxima lines that point to smoothed impulses. For each cluster centered in a 64 × 64 pixel subimage we calculated average general modulus maxima magnitude. The results of the calculation of general modulus maxima averages are shown in **Table 1**.

The local and total averages, presented in **Table 1**, are used to estimate capabilities of the applied method for the microcalcification detection. The average magnitude of general maxima of all detected eligible smoothed impulse functions in healthy areas was 3.59×10^{-5} (**Table 1**, Total-Column 6), while the average magnitude in areas with diagnosed microcalcification clusters was 4.18×10^{-5} (**Table 1**, Total-Column 5), which is 16% increase (**Table 1**, Total-Column 7). Similarly, the increase exists in each tested background tissue (20%, 18%, and 11% in the case of glandular, dense and fatty tissue, respectively) showing that microcalcification detection based on this feature is not qualitatively sensitive on background tissue.

The minimal computed average magnitude in an abnormal dense area was 3.49×10^{-5} (**Table 1**, Dense-Column 3), which was lower than corresponding average in normal dense areas 3.56×10^{-5}(**Table 1**, Dense-Column 6), showing that hard absolute thresholding (for example at 3.50×10^{-5}) would not result in proper classification. Although the microcalcification related additional criterions would help the proper classification, in this analysis we evaluate the magnitude of general maxima in microcalcification detection leaving the combination of criterions for further research.

As we said, the averages in Column 5 and Column 6 in **Table 1** could be considered in both linear and non-linear part of the diagrams presented in Figures 5-7. We conclude that a threshold based on the local average of gM magnitude will yield an appropriate classification of singular points. Further lowering the threshold would cause more false positive findings and therefore requires the additional classification criterions, while higher threshold would decrease classification sensitivity.

Table 1. Local and total gM magnitude averages in breast tissue with abnormal and normal findings.

Case		Abnormal			Normal	Rate
		Local Average		Total	Total	
		min ×10^{-3}	max ×10^{-3}	aver. ×10^{-3}	aver. ×10^{-3}	$\frac{C(5)}{C(6)}$
1	2	3	4	5	6	7
	Total (all)	3.49	5.47	4.18	3.59	1.16
Abnormality	Malignant	3.53	5.28	4.12	N/A	N/A
	Benign	3.49	5.47	4.23	N/A	N/A
Background Tissue	Glandular	3.63	5.47	4.21	3.50	1.20
	Dense	3.49	5.33	4.19	3.56	1.18
	Fatty	3.84	4.58	4.12	3.71	1.11

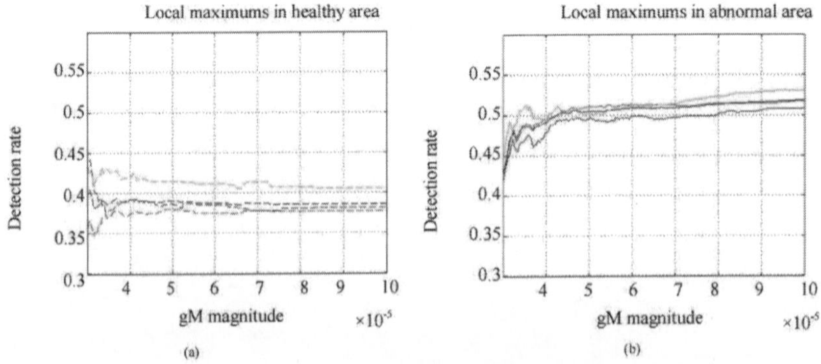

Figure 7. Smoothed impulse function with local intensity maximum in its neighborhood detection rate $DRF_{local\_max}(x)$ comparison for healthy tissue dashed lines (a) and the regions consisting diagnozed microcalcification clusters full lines (b). Black lines show the behavior of detection rate vs gM magnitude for all tested clusters, while blue, red, and green lines are related to glandular, dense, and fatty tissue, respectively.

4.3. Statistical Analysis of Singularities in Breast Tissue

In following analysis, a magnitude threshold of the general modulus maxima was determined as the local average of all gM magnitudes of all eligible maxima lines with detected negative Hölder exponent in the area 64 × 64 pixels centered in a diagnosed cluster. This choice of the adaptive threshold has reduced the number of dominant singularities down to 970 (31.1%) of all 3115 eligible singularities (**Table 2**, Total, Column 4), while 1179 (37.8%) were pointing either malignant or benign microcalcifications. The true positive fraction (sensitivity) was 65.2%, which results in ROC curve area index $A_z = 0.77$.

In the case of the malignant microcalcifications we find 33.3% as suspicious, while the malignant microcalcifications were 48.7%, meaning that the average threshold is too high. In **Table 3** we see that only 2% were false positive against 17.5% false negative findings. True positive fraction was 64.1%, while ROC curve area index was $A_z = 0.80$.

We find a different tendency in the case of benign microcalcification. The average threshold classified 29.1%, while benign microcalcifications were 27.3%, meaning that the average threshold was set a little lower than optimal. False positive and false negative findings were similar (10.7% and 9.0%, respectively). Surprisingly, the other statistical measures were not significantly different, i.e. true positive fraction was 67.1%, while ROC curve area index was $A_z = 0.76$. The details are shown in Tables 2 and 3.

Average gM magnitude of microcalcification was 4.58×10^{-5}, while the average of all eligible maxima lines was 4.18×10^{-5}, leaving additional room for selection improvement based on the magnitude of general maxima. Similar conclusion holds for separately analyzed both malignant and benign clusters where average magnitudes corresponding to microcalcifications were $4.34 \times 10^{-5}$and 5.02×10^{-5} respectively while threshold averages were 4.12×10^{-5} and 4.23×10^{-5}respectively. The difference of average magnitudes 4.34×10^{-5} and 5.02×10^{-5} for malignant and benign microcalcifications may lead to microcalcification malignancy probability estimation in further research.

4.4. Fractal Analysis of Singularities in Breast Tissue

In the following sub-sections, we determine fractal dimension of selected singularities for 3 types of data-all detected singularities that resemble smoothed impulse function, detected smoothed impulse functions with dominant general modulus maxima, and the detected singularities that correspond to singularities in diagnosed microcalcification clusters.

Partition functions $Z(q, a)$, $q = 0$ and 0.5 £ $\log_2(a)$ £ 1.5, of all detected singularities in 25 microcalcification clusters are presented in **Figure 8**. The rhombi represent all singularities in all 25 analyzed microcalcification clusters. The squares represent all singularities with diagnosed malignant abnormalities in 12 clusters, while the triangles represent all singularities detected in remaining 13 clusters.

Specifically the slopes of $\log_2(Z(0,a))$ vs. $\log_2(a)$, 0.5 £ $\log_2(a)$ £ 1.5, correspond to the fractal dimension of the sets of singularities for all detected singularities (rhombi), as well as for 12 malignant (squares) and 13 benign (triangles) clusters. The trend-lines show the slope is close to 2 in all three analyzed datasets. Notice that in 2-D space, fractal dimension can not be considered higher than 2. The results show that the eligible singularities are uniformly distributed everywhere in the analyzed mammogram sub-images.

Similarly, the slopes of $\log_2(Z(0,a))$ vs $\log_2(a)$, 0.5 £ $\log_2(a)$ £ 1.5, corresponding to selected dominant singularities, as well as diagnosed microcalcification clusters are determined and shown in**Table 4**.

In the case of microcalcification clusters, dominant and diagnosed singularities have similar fractal dimension values (between 1.4 and 1.5), which indirectly supports the assumption that dominant eligible singularities point to microcalcifications. There is no difference between malignant and benign cluster fractal dimensions in 200 mm/pixel spatial resolution. Notice that for microcalcification cluster classification purpose spatial resolution 200 µm/pixel doesn't allow zooming at 70 µm/pixel where microcalcification edge becomes visible and malignancy analysis by fractal dimension possible.

Dominant singularities in healthy areas have significantly lower fractal dimension, i.e. tending to 1 for the average of all 20 mammograms. Fractal dimension approaching 1 is the sign of linear structures typical for detected edges. Fractal analysis shows that smoother signal transition is uniformly distributed everywhere producing fractal dimension tending to 2 and 1.73 in normal and abnormal breast tissue respectively. On another hand fractal dimension of detected dominant signal transition in normal tissue is tending to 1 implying that the dominant singularities belong to the curve structures.

Healthy tissue fractal analysis shows that the set of all detected singularities has fractal dimension of 1.73 while the set of locally dominant singularities has fractal dimension tending to 1. This difference approves the significance of general modulus maxima in the detection of microcalcification. Fractal dimension of 1 is typical for the linear structures in the tissue and can be used in mammogram segmentation which we are proposing to be a future extension of this work.

Healthy tissue statistical analysis confirms our expectation that general maxima of normal fluctuation with negative Hölder exponent are lower than the general maxima belonging to the maxima lines representing the abnormalities suggesting that a combination of global and local thresholding need to be employed.

Table 2. Abnormality detection statistics in 25 analyzed clusters.

Abnormality	No. of eligible max. lines	No. of domin max. lines	Column (3)/(2)%	No. of accurate max. lines	Column (5)/(2)%
1	2	3	4	5	6
Total (all)	3115	970	31.1	1179	37.8
Total (malignant)	1533	510	33.3	747	48.7
Total (benign)	1582	460	29.1	432	27.3

Table 3. Microcalcification detection performance in 25 analyzed clusters.

	All clusters		Malign. clusters		Benign clusters	
	No. of eligible maxima lines	%	No. of eligible maxima lines	%	No. of eligible maxima lines	%
1	2	3	4	5	6	7
Total eligible	3115	100	1533	100	1582	100
True classified	2504	80.4	1234	80.5	1270	80.3
False classified	611	19.6	299	19.5	312	19.7
True positive	769	24.7	479	31.2	290	18.3
True negative	1735	55.7	755	49.2	980	61.9
False positive	201	6.5	31	2.0	170	10.7
False negative	410	13.2	268	17.5	142	9.0

Table 4. Fractal dimmension of singularities in microcalcification clusters and healthy tissue.

Tissue		Total	Malignant	Benign
1	2	3	4	5
Abnormal	Eligible	2.00	2.00	1.97
	Dominant	1.43	1.46	1.40
	Diagnosed	1.47	1.47	1.47
Normal	Eligible	1.73	N/A	N/A
	Dominant	1.04	N/A	N/A

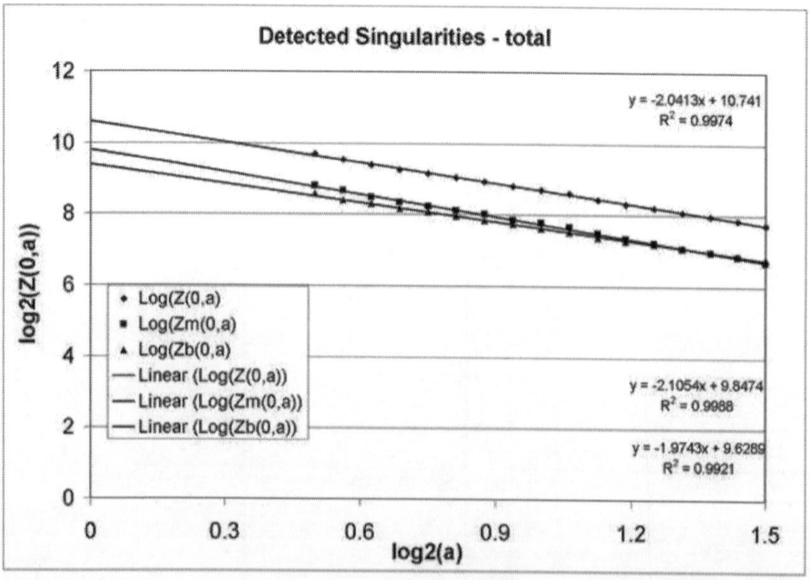

Figure 8. The fractal dimension for all detected singularities in the 25 analyzed microcalcification clusters in all, malignant, and benign sets of data (the slope of the partition function, q = 0 and a®0) is around 2, showing that the singular points are distributed everywhere in the clusters.

4.5. A Microcalcification Detection Simulation Example

Our proposed algorithm successfully localizes both malignant and benign microcalcification clusters and it is worth to notice that no other features but dominant general modulus maximum with local average threshold is employed in the detection of microcalcification clusters such as those shown in **Figure 9**.

Wang and Karayiannis [14] suggested detection of high frequency energy blobs with no differentiation between sharp transition characterized as smoothed impulse function and normal high frequency signal fluctuation in mammograms. Based on that idea we performed detection of energy blobs of detected dominant general modulus maxima. We consider that a microcalcification is a bright spot with a local intensity maximum in the neighborhood of a singular point related to the microcalcification and eliminate the singular point without a local intensity maximum in its neighborhood. The mammogram mdb223 [22] with two benign microcalcification clusters and a film damage is shown in **Figure 9**(a). The yelloworange-red iso-energy lines point to the suspicious spots with detected energy peaks as shown in **Figure 9**(b). Film damage was not disturbing the detection process, because of missing local maxima in the eligible neighborhood of the detected singularities in the border of film damage.

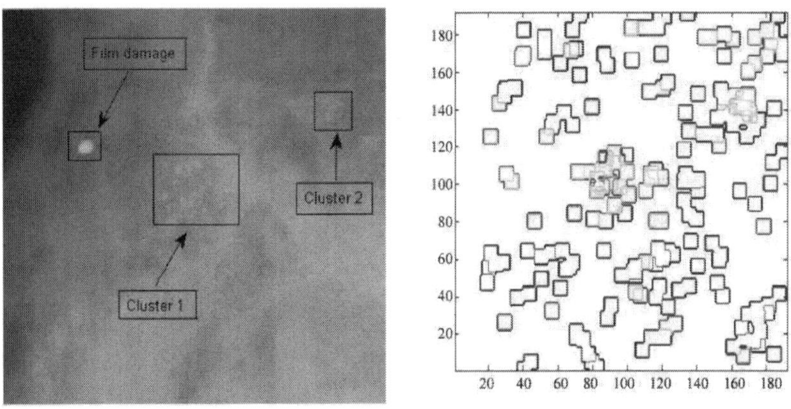

(a) (b)

Figure 9. Mammogram mdb223 [24] with two microcalcification clusters and a film damage (a), and detected significant energy peaks are represented by yellow-orange-red iso-energy lines (b).

5. CONCLUSIONS AND FUTURE WORK

Continuous wavelet transform was employed to detect singularities in mammograms by tracking modulus maxima along maxima lines across the scale and integrating it with fractal properties to differentiate between normal signal fluctuation and irregular ones in breast tissue.

Microcalcifications in mammograms were modeled by smoothed impulse function and detected by their negative Hölder exponent. Proposed procedure for maxima line classification enables selected general maxima to be analyzed by their magnitudes.

This method is capable to make difference between smooth signal transitions inevitable present in mammograms and microcalcification as sharper signal transition. Presented numerical results show that selecting maxima lines with negative Hölder exponent and local thresholding of general modulus maxima make possible the detection of the suspicious spots as local energy peaks of general modulus maxima.

Dominant general modulus maxima highly map into diagnosed microcalcifications in terms of local average magnitude and spatial distribution. Simulation results demonstrated that smoothed impulse function, recognized by their fractal properties of localized wavelet transform modulus maxima, can be considered a reliable feature for microcalcification detection in mammograms. Moreover, comparison of healthy and calcified breast tissue based on both statistical and fractal results show that selected general modulus maximum is a significant feature in microcalcification detection and classification.

Our results show that proposed metric for microcalcification detection is not sensitive on background tissue and therefore particularly convenient for dense parenchyma background in which microcalcification detection can be often a very challenging task.

Although a part of information has been lost by spatial resolution resampling from 50 mm to 200 mm microcalcifications preserve their smoothed impulse function nature, which has been detectable by proposed method. We showed that energy peaks in modulus maxima field correspond to microcalcification suspicious spots.

Coarse resolution makes faster suspicious spots searching procedure in real time during the screening procedure, which can be beneficiary by preventing an additional screening and saving the patients from having to endure the uncomfortable feelings during the procedure and more importantly, emotionally surviving cancer while waiting for additional test results.

Involving local intensity maximum of underlying mammogram in the neighborhood of detected smoothed impulse function significantly reduces false positive findings. Then the algorithm allows smaller threshold value of the general modulus maxima and therefore becomes more sensitive by detecting eligible smoothed impulse function with lower magnitude necessary for early detection of breast cancer.

The number of detected microcalcifications per area unit could be another feature that could reduce false positive findings.

Our findings in the difference of average general maxima magnitudes for malignant and benign microcalcifications leave a space for microcalcification malignancy probability estimation in further research. Namely, the numerical results show that general modulus maxima average of cancerous microcalcifications is lower than the average of benign ones. Lower modulus maximum magnitudes of cancerous findings make the detection more difficult, but open a space for improved classification in terms malignant vs. benign microcalcifications. Additional analysis might show that general maximum magnitude could serve as a parameter for malignancy probability estimation in real time.

The areas with microcalcification clusters tend to have higher fractal dimension than the healthy areas. Practical implementation of how to use fractal dimension in microcalcification detection and/or classification can be another interesting future work.

We showed that initial microcalcification segmentation could be performed based on wavelet transform modulus magnitude and phase and underlying mammogram in coarse resolution not significantly disturbed by noise. If necessary the microcalcification segmentation could be then refined in the finer resolution in which image acquisition was performed.

We found that our implementation of this proposed work including all parameter values is appropriate for the 200 µm/pixel spatial resolution and 8 bits/pixel numerical resolution. This approach could be easily applicable to other mammogram databases with different spatial and numerical resolution if proper adjustment to the estimates of the parameters values were taken into consideration.

Finally, involving additional microcalcification features could decrease false positive findings by enabling lowering modulus maximum threshold, which improves detection sensitivity of applied algorithm and therefore the general modulus maximum of smoothed impulse function can be used not only independently but also integrated with other microcalcification detection algorithms.

REFERENCES

1. S. Mallat and W. L. Hwang, "Singularity Detection and Processing with Wavelets," IEEE Transactions on Information Theory, Vol. 38, No. 2, 1992, pp. 617-643.

2. S. Mallat and S. Zhong, "Characterization of Signals from Multiscale Edges," IEEE Transactions on Pattern Analysis and Machine Intelligence, Vol. 14, No. 7, 1992, pp. 710-732.

3. S. Mallat, "A Wavelet Tour of Signal Processing," Academic Press, Waltham, 1999.

4. Y. Meyer, "Un Contre-Exemple a la Conjecture de Marr et Celle de S. Mallat," Preprint, 1991.

5. E. B. Arneodo and J. F. Muzy, "The Thermodynamics of Fractals Revisited with Wavelets," Physica A, Vol. 213, 1995, pp. 232-275.

6. A. Arneodo, B. Audit, E. Bacry, S. Manneville, J. F. Muzy and S. G. Roux, "Thermodynamics of Fractal Signals Based on Wavelet Analysis: Application to Fully Developed Turbulence Data and DNA Sequences," Physica A, Vol. 254, 1998, pp. 24-45.

7. N. Decoster, "Analyse Multifractale d'Images de Surfaces Rigueuses à l'aide da la Transformation en Ondolettes," Ph.D. Thesis, L'Université de Bordeaux I, 1999. http://decoster.free.fr/phd

8. A. Arneodo, N. Decoster and S. G. Roux, "A WaveletBased Method for Multifractal Image Analysis. I. Methodology and Test Applications on Isotropic and Anisotropic Random Rough Surfaces," The European Physical Journal B, Vol. 15, 2000, pp. 567-600.

9. S. Roux, J. F. Muzzy and A Arneodo, "Detecting Vorticity Filaments Using Wavelet Analysis: About the Statistical Contribution of Vorticity Filaments to Intermittency in Swirling Turbulent Flows," The European Physical Journal B, Condensed Matter Physics, Vol. 8, No. 22, 1999, pp. 301-322.

10. J. J. Heine and P. Malhotra, "Mammographic Tissue, Breast Cancer Risk, Serial Image Analysis, and Digital Mammography. Part 1. Tissue and

Related Risk Factors," Academic Radiology, Vol. 9, No. 3, 2002, pp. 298-316.

11. J. J. Heine and P. Malhotra, "Mammographic Tissue, Breast Cancer Risk, Serial Image Analysis, and Digital Mammography. Part 2. Serial Breast Tissue Change and Related Temporal Influences," Academic Radiology, Vol. 9, No. 3, 2002, pp. 317-335.

12. M. Kallergi, "Computer-Aided Diagnosis of Mammographic Microcalcification Clusters," Medical Physics, Vol. 31, No. 2, 2004, pp. 314-326.

13. J. J. Heine, S. R. Deans, D. K. Cullers, R. Stauduhar and L. P. Clarke, "Multiresolution Statistical Analysis of HighResolution Digital Mammograms," IEEE Transaction on Medical Imaging, Vol. 16, No. 5, 1997, pp. 503-515.

14. T. C. Wang and N. B. Karayiannis, "Detection of Microcalcifications in Digital Mammograms Using Wavelets," IEEE Transactions on Medical Imaging, Vol. 17, No. 4, 1998, pp. 498-509.

15. W. Zhang, H. Yoshida, R. M. Nishikawa and K. Doi, "Optimally Weighted Wavelet Transform Based on Supervised Training for Detection of Microcalcifications in Digital Mammograms," Medical Physics, Vol. 25, No. 6, 1998, pp. 949-956.

16. R. N. Strickland and H. I. Hahn, "Wavelet Transforms for Detecting Microcalcifications in Mammograms," IEEE Transactions on Medical Imaging, Vol. 15, No. 2, 1996, pp. 218-229.

17. F. J. Gilbert, S. M. Astley, M. A. McGee, M. G. Gillan, C. R. Boggis, P. M. Griffiths and S. W. Duffy, "Single Reading with Computer-Aided Detection and Double Reading of Screening Mammograms in the United Kingdom National Breast Screening Program," Radiology, Vol. 241, No. 1, 2006, pp. 47-53.

18. P. Taylor, J. Champness, R. Given-Wilson, K. Johnston, and H. Potts, "Impact of Computer-Aided Detection Prompts on the Sensitivity and Specificity of Screening Mammography," Health Technology Assessment, Vol. 9, No. 3, 2005, pp. 1-58.

19. J. J. Fenton, S. H. Taplin, P. A. Carney, L. Abraham, E. A. Sickles, C. D'Orsi, E. A. Berns, G. Cutter, R. E. Hendrick, W. E. Barlow and J. G. Elmore, "Influence of ComputerAided Detection on Performance of Screening Mammography," The New England Journal of Medicine, Vol. 356, 2007, pp. 1399-409.

20. L. E. Philpotts, "Can Computer-Aided Detection Be Detrimental to Mammographic Interpretation?" Radiology, Vol. 253, No. 1, 2009, pp. 17-22.

21. J. Tang, R. M. Rangayyan, J. Xu, I. E. Naqa and Y. Yang, "Computer-Aided Detection and Diagnosis of Breast Cancer with Mammography: Recent Advances," IEEE Transactions on Information Technology in Biomedicine, Vol. 13, No. 2, 2009, pp. 236-251.

22. J. Suckling, "The Mammographic Image Analysis Society Digital Mammogram Database," Exerpta Medica International Congress, Vol. 1069, 1994, pp. 375-378. http://peipa.essex.ac.uk/info/mias.html

23. B. Mandelbrot, "How Long Is the Coast of Britain? Statistical Self-Similarity and Fractional Dimension," Science, New Series, Vol. 156, No. 3775, 1967, pp. 636-638.

24. T. Netsch and H. O. Peitgen, "Scale-Space Signatures for the Detection of Clustered Microcalcifications in Digital Mammograms," IEEE Transactions on Medical Imaging, Vol. 18, No. 9, 1999, pp. 774-786.

Chapter 7

Dose Reduction by the Use of a Wavelet-Based Denoising Method for Digital Radiography

Haruyuki Watanabe[1], Du-Yih Tsai[2], Yongbum Lee[2], Eri Matsuyama[3]

[1]School of Radiological Technology, Gunma Prefectural College of Health Sciences, Gunma, Japan

[2]Department of Radiological Technology, Graduate School of Health Sciences, Niigata University, Niigata, Japan

[3]Department of Radiological Technology, Faculty of Fukuoka Medical Technology, Teikyo University, Fukuoka, Japan

ABSTRACT

The primary purpose of this paper is to provide a novel wavelet-domain method for digital radiography with low dose examination. Approach of this study is an improved wavelet-transform- based method for potentially reducing radiation dose while maintaining clinically acceptable image quality. The wavelet algorithm integrates the advantages of wavelet-coefficient-weighted method and the existing Bayes Shrink thresholding method. In order to confirm the usefulness of the proposed method, the resolving and noise characteristics of the processed computed radiography images were measured. In addition, variations of contrast and noise with respect to radiation dose were also examined. Finally, to verify the effect of clinical examination, visual evaluations were also performed in lower abdominal area using phantom. Our quantitative results demonstrated that our wavelet algorithm could improve resolution characteristics while keeping the noise level within acceptable limits. Visual evaluation result demonstrated that the proposed method was superior to other published methods. Our proposed method recognized effect on decreasing in exposure dose in lower abdominal radiographs. As a conclusion, our proposed method's performance is better when compared with that of the 3 conventional methods. The proposed method has the potential to improve visibility in radiographs when a lower radiation dose is applied.

KEYWORDS

Radiation Dose, Image Quality, Image Processing, Noise Reduction, Wavelet Transforms

1. INTRODUCTION

The issue of radiation dose exposure to patients from digital radiography is a major public health concern. In particular, it is important to keep radiation dose exposure to a minimum in female patients during their reproductive period, who frequently undergo repeated radiation exposure during the course of diagnostic imaging and treatment follow-up.

It is known that a trade-off exists between noise level and radiation dose. On the one hand, high-dose radiation will lower the noise level, but may expose the patient to excessive radiation. On the other hand, low-dose radiation will lower the signal-to-noise ratio of the image and result in reducing the amount of image information. The balancing of radiation dose and image quality should be performed precisely to ensure that patient doses are kept at a reasonable minimum, while maintaining clinically acceptable image quality. To address this issue, much research, including the development of new detectors and image processing methods [1] -[3] has been carried out. In recent years, several investigators have reported that wavelet-based image processing techniques are effective in the reduction of radiation dose [4] -[13] .

Conventional radiography is widely used for the pelvis and lumbar spine. However, the radiation dose for pelvic and lumbar x-ray examinations using a radiograph is relatively high in order to obtain acceptable image quality. An effort to reduce the exposure dose can have a positive effect on a patient's quality of life.

In this study, we propose an improved wavelet-transform-based method for potentially reducing the radiation dose while maintaining clinically acceptable image quality. The proposed method integrates the advantages of our previously proposed wavelet-coefficient-weighted method [11] -[13] and the existing Bayes Shrink thresholding method [14] . The wavelet-coefficient-weighted method has the advantage of effective edge enhancement accompanied by a slight suppression of noise increase (however, the noise is also enhanced definitely). In contrast, the Bayes Shrink thresholding method [14] is used for denoising, while the edge is preserved as much as possible (however, the edge also decreases definitely). It is expected that our approach, integrating the two methods, can achieve edge enhancement with almost no significant noise increase and can be applied to low- dose radiographs to improve image quality. To verify the proposed method's effectiveness in reducing radiation dose in digital radiography, it was assessed quantitatively and qualitatively.

2. METHODS

2.1. Proposed Method

The main steps of the proposed method include a previously reported wavelet coefficient adjustment technique for contrast enhancement [11] - [13] and a wavelet thresholding technique for noise reduction. Figure 1 shows a schematic diagram of the proposed method. As shown inFigure 1, the proposed method for denoising radio graphic images starts by decomposition of the original image by use of the discrete wavelet transform, which results in obtaining different detail wavelet coefficients (horizontal, vertical, diagonal). The three detail coefficients are then processed by use of a sigmoid-type transfer curve for adjustment of wavelet coefficient, followed by Bayes Shrink thresholding.

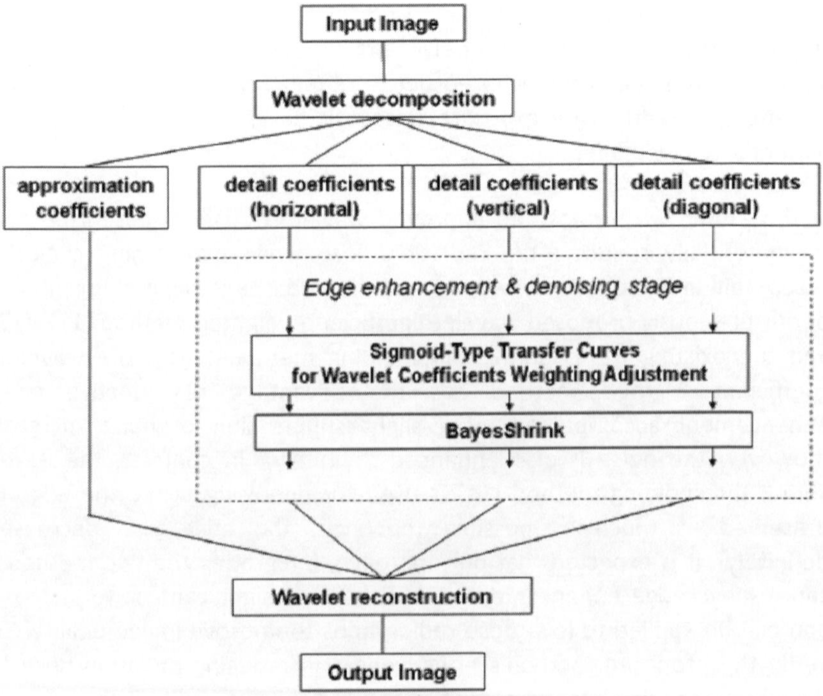

Figure 1. Flow chart of the proposed method.

2.1.1. Sigmoid-Type Transfer Curve for Wavelet Coefficient Weighting Adjustment

A sigmoid-type transfer curve with a one-to-one mapping function is used for enhancement of image contrast. The mapping function was determined based on the following considerations: a) in the case of detail compo- nents at a specific level, high-value coefficients are weighted because they carry effective information; b) the coefficients at low levels are heavily weighted, because they carry detailed information, such as edge information; and c) the approximation coefficients are not manipulated in order to prevent image distortion.

In the wavelet decomposition of level j, the sigmoid-type transfer curves of wavelet coefficients can be expressed as follows:

$$w^{j}_{\text{output}}(m,n) = a_0 \times \frac{1}{1 + 1 \Big/ \exp\left[\frac{\left|w^{j}_{\text{input}}(m,n) - c_0\right|}{b_0}\right]} \times w^{j}_{\text{input}}(m,n), \tag{1}$$

where $w^{j}_{\text{input}}(m,n)$ and $w^{j}_{\text{output}}(m,n)$ represent input and output coefficients, respectively. a_0, b_0, and c_0 are constants. In practice, Equation (2) is used as the mapping function instead of Equation (1). In Equation (2), the values of the coefficients are expressed in terms of percentage for the ease of computation.

$$v^{j}_{\text{output}}(m,n) = a \times \frac{1}{1 + 1 \Big/ \exp\left[\frac{v^{j}_{\text{input}}(m,n) - c}{b}\right]} \times v^{j}_{\text{input}}(m,n), \tag{2}$$

Here, $w^{j}_{\text{input}}(m,n)$ is the input value expressed in terms of percentage. This value makes the mean of the absolute values of the coefficients at level j equal to 50%. $w^{j}_{\text{output}}(m,n)$ is the corresponding output value, also expressed in terms of percentage. The reason for utilization of percentage is that by doing so the constants a, b, and c could be used independent of image characteristics. The constants a, b, and c are determined depending on the extent of enhancement. In this study, the value of a was computed by use of Equation (3):

$$a = 2 - \frac{(j-1)}{N}, \tag{3}$$

where N represents the maximum decomposition level. Consequently, the lower the wavelet decomposition level j, the greater the gradient of the transfer curve becomes. The reason for making the value of a dependent on decomposition levels is that the wavelet coefficients at low-decomposition levels, which mainly contain information about edges, are highly weighted. The constant c was determined by use of Equation (4):

$$c = d + b \times \ln(a - 1.0), \tag{4}$$

where d is a constant used for determining the inflection point of the sigmoid curve, and b represents a constant used for determining the gradient of the sigmoid curve. The values of b and d used in this study were 20 and 25, respectively [12] . The values were determined empirically.

2.1.2. Bayes Shrink Thresholding Technique

The wavelet transform, due to its excellent localization property, has become an indispensable image-denoising tool during the past two decades. Wavelet de noising attempts to remove the noise present in an image while preserving the image characteristics. Wavelet thresholding, first proposed by Donoho [4] , is a signal-estimation technique that exploits the capabilities of the wavelet transform for signal and image denoising. It removes noise by eliminating coefficients that are insignificant relative to some threshold. Therefore, the selection of the threshold is the most important step in wavelet-based denoising techniques.

Various threshold selection methods have been proposed, such as Visu Shrink [15] , Sure Shrink [16] , and Bayes Shrink [14] . In the Visu Shrink method, a universal threshold that is a function of the noise variance and the number of samples is developed based on the minimax error measure. The threshold value in the Sure Shrink method is optimal in terms of the Stein's unbiased risk estimator. The Bayes Shrink method determines the threshold value in a Bayesian framework, through modeling of the distribution of the wavelet coefficients as Gaussian [17] . Several researchers have compared the three selection methods, and their results have shown that Bayes Shrink outperforms the other two methods [17] - [19] . In this study, we employed the Bayes Shrink method for denoising.

In the Bayes Shrink scheme, the threshold is determined for each sub and by assuming a generalized Gaussian distribution (GGD) [14]. The Bayes Shrink threshold, T_B, is given as

$$T_B = \frac{\sigma^2}{\sigma_X}, \tag{5}$$

where σ^2 is the noise variance and σ_X the signal standard deviation. Suppose that the signal and the noise are independent of each other. The estimated variance of the observed image, σ_Y^2, can be written as

$$\sigma_Y^2 = \sigma_X^2 + \sigma^2, \quad (6)$$

The estimated variance of signal σ_X^2 is then deduced by

$$\sigma_X = \sqrt{\max\left(\sigma_Y^2 - \sigma^2, 0\right)}, \quad (7)$$

A robust estimator of the noise variance is obtained by

$$\sigma = \frac{M}{0.6745}, \quad (8)$$

where M is the median value of the absolute wavelet coefficients in the first decomposition level. A detailed explanation of the Bayes Shrink method is given in [14].

2.2. Data Acquisition

Images that were used for measurement of physical characteristics were acquired with use of a multipurpose phantom [20] . Figure 2 shows an example of phantom images. A computed radiography (CR) system (FCR XG-1, Fuji Photo Film, Tokyo, Japan) and an imaging plate (ST-V_N, Fuji Photo Film, Tokyo, Japan) were used in this study. A pixel size of 0.1 mm and a quantization level of 10 bits were employed for data acquisition. The system parameter settings for the latitude (L) and sensitivity (S) were fixed at 3 and 200, respectively. Images were taken with a radiation quality of RQA-5 (HVL = 7.1 mm Al, 21 mm Al additional filtration) by using a tungsten target x-ray tube (Hitachi, Tokyo, Japan). The focal spot size of the x-ray tube was 0.6 mm. The source- to-image receptor distance was 190 cm. The amount of exposure was 4.63×10^{-7} C/kg (50 mAs). Twenty phantom images were obtained and used for measuring the presampled modulation transfer function (MTF), noise power spectrum (NPS), and gray level contrast (GLC).

Four different radiation levels were used for investigating the effect of the physical characteristics on the radiation dose. The 4 radiation level ratios with respect to the reference level, 4.63×10^{-7}C/kg, were 50/100, 64/ 100, 80/100, and 100/100.

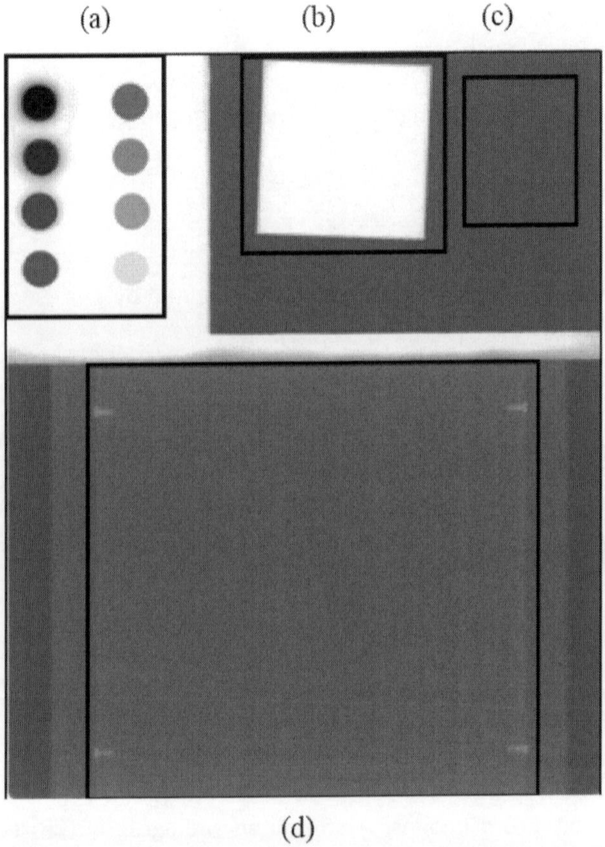

Figure 2. Examples of phantom images used for the measurement of physical characteristics. (a) Contrast detail curve; (b) MTF; (c) NPS; and (d) GLC using acrylic disk on Burger phantom.

A visual evaluation of wavelet-processed images of a human body phantom was performed to confirm the effectiveness of the proposed method in reducing radiation dose. An anterior-posterior (AP) projection of the hip joint and a lateral view of the lumbar spine on the human body phantom were exposed to various dose levels. These two images were also taken at four different radiation level ratios, 50/100, 64/100, 80/100, and 100/100, instead of the reference level that is commonly used in clinical radiology practice. In this study, the hip joint phantom was exposed at 70 kVp and 32 mAs, and the lumbar phantom at 82 kV and 64 mAs.

2.3. Measurement of Physical Characteristics

2.3.1. MTF

The presampled MTFs were measured with an angled-edge method [21] . The edge device was made of a 1-mm- thick sharp-edged tungsten plate whose dimensions were 10 × 10 cm^2. The direction of the edge was oriented with a small angle (2° - 3°). The edge spread function (ESF) in the direction perpendicular to the edge was then obtained. To reduce the noise in the edge profile, 20 representations of the sampled ESFs were generated from the region of interest (ROI). Then the ESFs were differentiated to obtain the line spread functions (LSFs), and the presampled MTFs were deduced by applying Fourier transformation to the LSFs [21] [22] . The resulting MTF was obtained by averaging the 20 MTFs.

2.3.2. NPS

NPS measurements were made by exposure of the imaging plate to a uniform beam of radiation. For determination of the NPS, a two-dimensional 2nd-order polynomial was fitted and subtracted to remove background trends. For the calculation, the central portion of each uniform image obtained was divided into 4 non-overlap- ping regions, 256 × 256 in size. A total of 80 regions were used. The NPS was calculated by applying the fast Fourier transform to each of the ROIs and then averaging the resulting spectrum estimates [23] [24].

2.3.3. GLC

A commercially available Burger phantom (Kyoto Kagaku, Kyoto, Japan) was employed for measurement of GLC characteristics. In this study, the GLC was used to describe the relative contrast of an image, defined by

$$GLC = \frac{\left| L_{acrylic} - L_{BG} \right|}{L_D - 1}, \quad (9)$$

where $L_{acrylic}$, L_{BG}, and L_D represent the mean pixel value of an 8.0 mm diameter circle of an acrylic disk 8.0 mm in thickness, the mean pixel value of the background, and the gray level of the CR, respectively. The GLC ranged from 0 to 1.0. Image contrasts with different gray levels could be compared because the GLC was normalized by (L_D ? 1). Low GLC corresponded to low contrast, while high GLC corresponded to high contrast. For clarifying the effect of the radiation dose on the GLC, dose ratios ranging between 100/100 and 50/100, instead of the standard dose, were measured.

2.4. Performance Comparison

In order to validate the superiority and effectiveness of the proposed method, the proposed method and 3 conventional methods, namely, the Wiener filter (WF) [25] [26] , Bayes Shrink method [14] , and sigmoid-type method [11] were compared. Table 1 shows various parameters of the three methods and the proposed method. The

WF used in this study was a pixelwise adaptive WF given as

$$g(i,j)=\mu+\frac{\sigma^2-v^2}{\sigma^2}\times\left(f(i,j)-\mu\right), \tag{10}$$

where f(i, j) and g(i, j) are the pixel values of pixel (i, j) in the input image and the output image, respectively. μ and σ^2 are the local mean and local variance, respectively, in the N-by-M local neighborhood of pixel (i, j) in the input image. M × N is the kernel size of the local region (a 5 × 5 matrix was used in this study as shown in Table 1). v^2 is the noise variance, which is the average of all the local variances in this study. Where the local image variance was large, the WF performed somewhat smoothly. Where the local image variance was small, the WF performed even more smoothly.

The proposed method and the above-described three methods were applied to the original images for performance comparison.

Table 1. List of methods and their parameters.

Method	Parameters
Wiener Filter (WF)	kernel Size = 5 × 5
Bayes Shrink (Bay)	Daubechies, Order = 4, Level = 4
Sigmoid (Sig)	Daubechies, Order = 4, Level = 4
Proposed (Pro)	Daubechies, Order = 4, Level = 4

2.5. Visual Evaluations

A visual evaluation was conducted by five experienced radiological technologists. The images were displayed on a liquid-crystal display (1280 × 1024 matrix, LCD-1980SXi, Nippon Electric Company, Tokyo, Japan). The parameters of window level, window width, and display image size on the image display apparatus were fixed. Each observer reviewed the images independently. The reading time was not limited. The five radiological technologists independently evaluated the total depiction of each phantom image for diagnostic acceptability by using a 5-point grading scale (1, no depiction; 2, faint; 3, acceptable; 4, good; 5, excellent). Statistical analyses were performed with the Friedman test. When a statistically significant difference was found ($p < 0.01$) in the five images (the original and the four image-processed images) at each dose ratio, pairwise comparisons were performed with Scheffe's method. Comparisons were made by use of five possible combinations, namely, WF-processed image, Bayes Shrink-processed image, sigmoid-processed image, proposed filter-processed image, and the original image.

3. RESULTS

Figure 3(a) shows the MTF values for the original image and the four processed images. The MTF value for the sigmoid image was the highest, followed by that for the proposed image. Both MTFs were considerably superior to the original image over the entire spatial frequency range. In contrast, the MTF values obtained from the Bayes Shrink and the WF images were slightly lower than that of the original image. Figure 3(b) shows the NPS values. The NPS values for the sigmoid image were pronouncedly higher than those of the original image. The NPS values for the proposed image were slightly higher or similar to those of the original image. In contrast, the NPS values for the Bayes Shrink and WF images were lower than those for the original image.

Figure 4(a) shows the GLC as a function of the radiation dose ratio. There were no significant differences in any of the GLCs. Figure 4(b) and Figure 4(c), respectively, show the NPSs as a function of the dose ratio at spatial frequencies of 1 mm^{-1} and 4 mm^{-1} for the original image and the four processed images. Although the trend of the values measured from the proposed-method-processed image is similar to that measured from the original image in dose ratios ranging from 80/100 to 50/100, the NPSs for the proposed method showed improvement in noise level at the dose ratio of 50/100 at the spatial frequency of 4 mm^{-1}.

Figure 5 illustrates the mean grades of visual evaluation for the hip joint and the lumbar spine at the four radiation dose ratios. In all cases, significant differences ($p < 0.01$) were found with the Friedman test at various radiation dose ratios. For the hip joint, the mean grade of the proposed method reached 3 points and was higher than those of the other methods tested at all radiation dose ratios. In the case of the lumbar spine, except for the radiation dose ratio of 50/100, the mean grade for the proposed method was higher than those of the other methods tested.

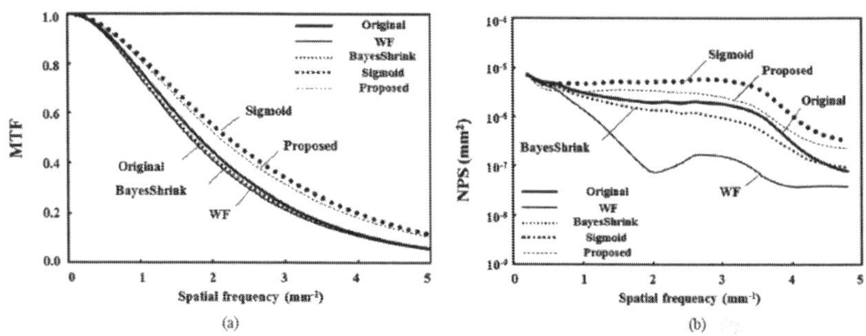

Figure 3. (a) Presampled MTFs for the original image and the four processed images; (b) NPSs for the original image and the four processed images.

Figure 6 illustrates visual evaluation results for the hip joint (AP) and lumbar spine (lateral) at various radiation dose ratios by use of Scheffe's method of paired comparisons. As shown in Figure 6, original image (Org) at 100/100 radiation dose ratio is located at the center (marked as zero) of the straight horizontal bar. The quality of the processed image was superior to that of the Org, if it had a higher score than the Org and there was a statistical significance. The quality of the processed image was inferior to that of the Org, if it had a lower score than the Org and there was a statistical significance. If the processed image had a similar score to that of the Org, the quality was considered to be equivalent to that of the Org. In terms of diagnostic acceptability, the proposed method provided significantly better results than those of the original image up to a 64/100 radiation dose ratio in the hip joint. When the radiation dose ratio was 50/100, no significant difference was found between the image processed by the proposed method and the original image.

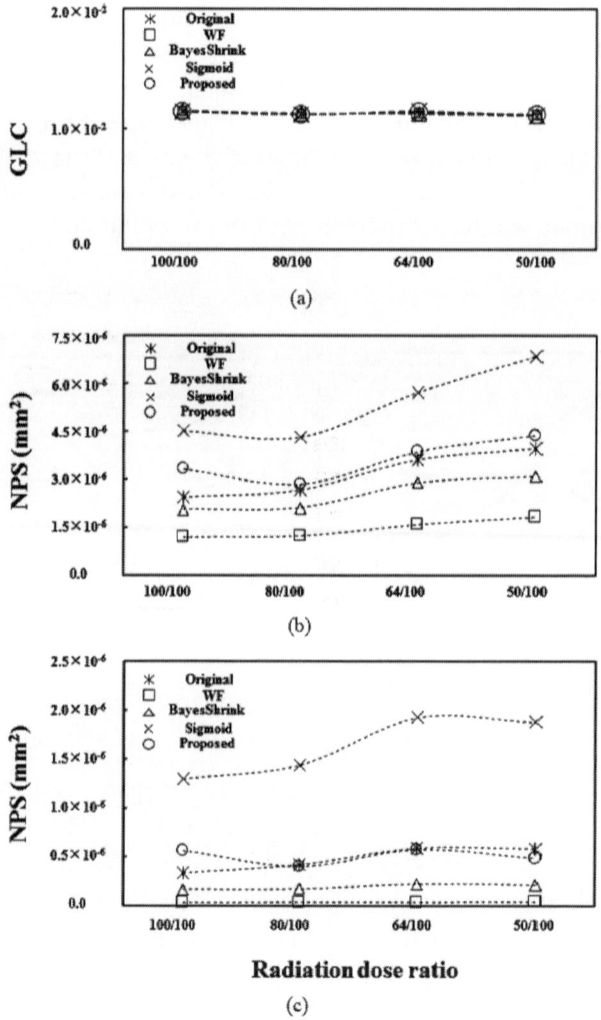

Figure 4. (a) Gray-level contrasts as a function of the dose ratio for the original image and the four processed images obtained using a Burger phantom. (b) The NPSs as a function of the radiation dose ratio, measured from the original image and the various images processed by the WF, BayesShrink, sigmoid-type, and the proposed methods at spatial frequency of 1 mm^{-1}. (c) The NPSs at spatial frequency of 4 mm^{-1}.

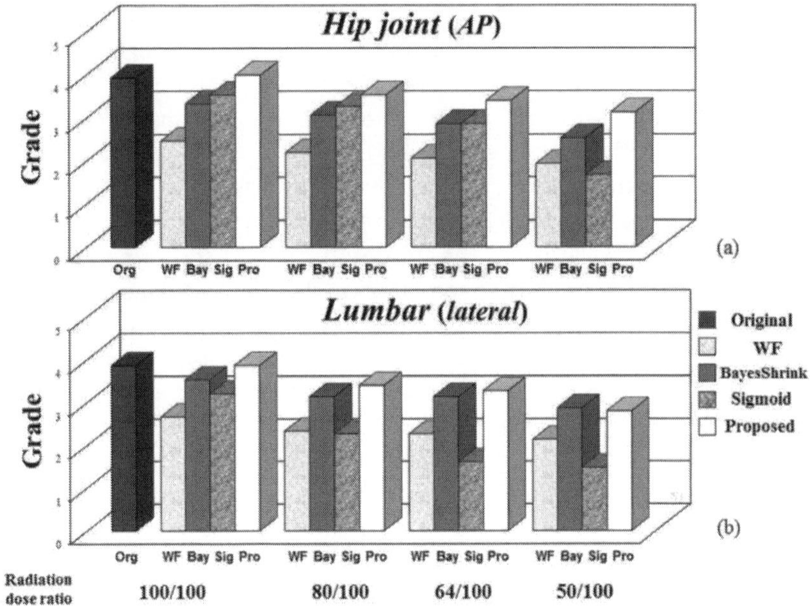

Figure 5. Mean grades of visual evaluation of the original and the four processed images for the hip joint and lumbar spine at the four radiation dose ratios, 100/100, 80/100, 64/100, and 50/100, in comparison to the standard dose. (a) hip joint (AP). (b) lumbar spine (lateral).

In the case of lumbar radiographs, the results obtained from the proposed method were comparable to those of the original image up to a 64/100 radiation dose ratio. However, the proposed method tended to show unsatisfactory results for a 50/100 radiation dose ratio.

Figure 7 illustrates original and processed images of the hip joint and lumbar spine of the human body phantom, which were used for the visual evaluation.

4. DISCUSSION

The proposed method provides the benefits of improved resolution and noise suppression. The experimental results demonstrated the method's effectiveness in dose reduction without degradation in image quality at a lower dose as compared to the standard dose. In the MTF and NPS measurements (Figure 3), the physical properties of the images processed by use of the sigmoid function and Bayes Shrink methods show distinct differences. The sigmoid function yields improved spatial resolution characteristics with increasing noise. In contrast, the Bayes Shrink method provides improved noise properties, but deteriorating spatial resolution. The proposed method incorporates the sigmoid method into the Bayes Shrink algorithm. As a result, the proposed method shows better spatial resolution and noise properties compared to the original image.

In the GLC measurements (Figure 4(a)), there were almost no differences in any of the GLCs. This implies that contrast was independent of x-ray dose and the proposed method did not contribute to contrast enhancement in the GLC experiment. This result was expected because a thin acrylic phantom was used in the experiment and CR has the characteristic of a linear relationship between pixel value and x-ray dose. However, the contrast of signal edge in a low-dose image of a thick scatterer, such as the human body, will decrease, because low-dose radiation brings about an increase in blur and noise. In this study, the proposed method contributed to the contrast enhancement of signal edge. Our previous study [11] indicated the potential for the contrast enhancement of signal edge by measuring the contrast improvement ratio for chest radiographs, mammograms, and chest CT images. We will investigate this issue in detail in future research.

In Figure 4(b) and Figure 4(c), the NPS values of our proposed method were near to those of the original image at all radiation dose ratios except 100/100. The high NPS value of our method at the 100/100 radiation dose ratio might be due to the fact that some noise enhanced by the sigmoid-function method was not recognized as noise in the Bayes Shrink method, and thus did not decrease efficiently.

As shown in Figure 5 and Figure 6, the results of our study indicate that the proposed strategy significantly improves the quality of low-dose images such that CR images obtained at 50% and 64% of the standard dose level provide clear depiction in AP views of the hip joint and in lateral views of the lumbar spine, respectively, in terms of visual evaluation. In Figure 7, the visibility of the overall appearance of bones seems to be improved by the proposed method. This may be due to the improvement in resolution and the suppression of noise. Maintaining a well-balanced relationship among contrast, spatial resolution, and noise is important. From this point of view, the proposed method has a well-balanced filter for the AP view of the hip joint and the lateral view of the lumbar spine at a lower dose.

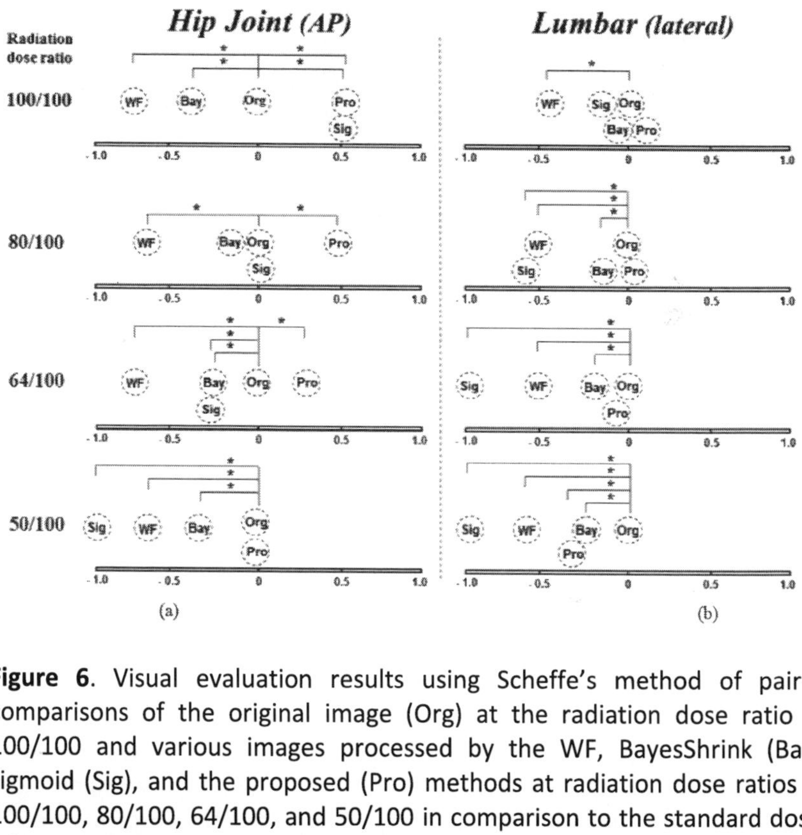

Figure 6. Visual evaluation results using Scheffe's method of paired comparisons of the original image (Org) at the radiation dose ratio of 100/100 and various images processed by the WF, BayesShrink (Bay), sigmoid (Sig), and the proposed (Pro) methods at radiation dose ratios of 100/100, 80/100, 64/100, and 50/100 in comparison to the standard dose. (a) Hip joint (AP). (b) Lumbar spine (lateral). There was a significant difference ($p < 0.01$) between the original image and the processed image at various dose ratios if the * mark is shown.

This study does have a limitation. The superiority of the proposed method was not demonstrated for the lateral lumbar spine when the radiation dose ratio was at 64/100 as compared to the standard dose. This issue may be overcome by optimizing parameters, such as b and d in Equation (4). To address this issue, further improvement of the proposed method is still needed.

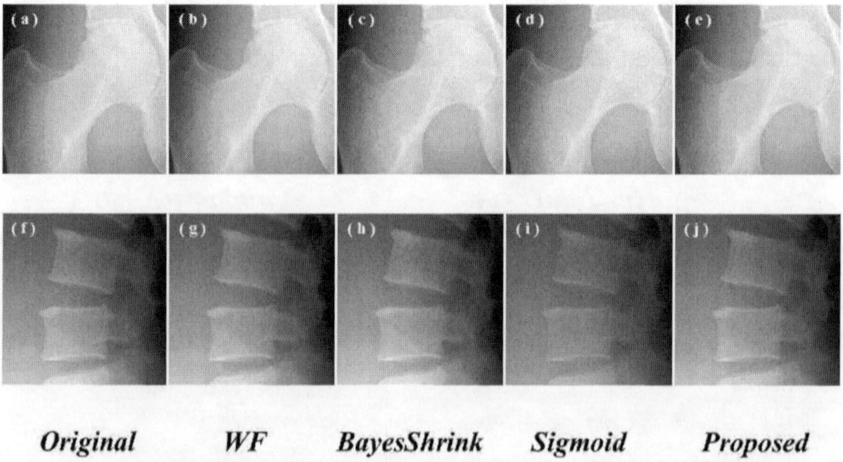

Figure 7. Original phantom images and processed phantom images obtained using WF, Bayes Shrink, sigmoid, and the proposed methods. (a)-(e) Hip joint at the standard dose; (f)-(j) Lumbar spine at the standard dose.

5. CONCLUSION

In this paper, we proposed an improved wavelet-transform-based method for potentially reducing radiation dose while maintaining clinically acceptable image quality. The effectiveness of the proposed method was demonstrated quantitatively and qualitatively. The experimental results showed that the proposed method could improve resolution while keeping noise level within acceptable limits. Furthermore, the results validated the effectiveness of our proposed method in the reduction of radiation dose. Our visual evaluation showed that an approximately 40% - 50% reduction in the exposure dose might be achieved with the proposed method in AP views of hip joint radiographs and lateral views of lumbar spine radiographs. The proposed method has the potential to improve visibility in radiographs when a lower radiation dose is applied.

ACKNOWLEDGEMENTS

Haruyuki Watanabe, Du-Yih Tsai, Yongbum Lee, Eri Matsuyama The authors also would like to thank the observers for their participation in visual evaluation.

REFERENCES

1. Gruber, M., Weber, M., Homolka, P., Nemec, S., Fruehwald-Pallamar, J. and Uffmann, M. (2011) Feasibility of Dose Reduction Using Needle-Structured Image Plates versus Powder-Structured Plates for Computed Radiography of the Knee. American Journal of Roentgenology, 197, 318-323.

2. Liu, X., Shaw, C.C., Lai, C.J. and Wang, T. (2011) Comparison of Scatter Rejection and Low-Contrast Performance of Scan Equalization Digital Radiography (SEDR), Slot-Scan Digital Radiography, and Full-Field Digital Radiography Systems for Chest Phantom Imaging. Medical Physics, 38, 23-33.

3. Schaefer-Prokop, C., Neitzel, U., Venema, H.W., Uffmann, M. and Prokop, M. (2008) Digital Chest Radiography: An Update on Modern Technology, Dose Containment and Control of Image Quality. European Journal of Radiology, 18, 1818-1830.

4. Donoho, D.L. (1995) De-Noising by Soft-Thresholding. IEEE Transactions on Information Theory, 41, 613-627.

5. Jansen, M., Uytterhoeven, G. and Bultheel, A. (1999) Image De-Noising by Integer Wavelet Transforms and Generalized Cross Validation. Medical Physics, 26, 622-630.

6. Harpan, M.D. (1999) A Computer Simulation of Wavelet Noise Reduction in Computed Tomography. Medical Physics, 26, 1600-1606.

7. Okamoto, T., Furui, S., Ichiji, H., Yoshino, S., Lu, J. and Yanagi, T. (2004) Noise Reduction in Digital Radiography Using Wavelet Packet Based on Noise Characteristics. Journal of Signal Processing, 8, 485-494.

8. Tischenko, O., Hoeschen, C. and Buhr, E. (2005) Reduction of Anatomical Noise in Medical X-Ray Images. Radiation Protection Dosimetry, 114, 69-74.

9. Ferrari, R.J. and Winsor, R. (2005) Digital Radiographic Image Denoising via Wavelet-Based Hidden Markov Model Estimation. Journal of Digital Imaging, 18, 154-167.

10. Yasuda, N., Ishikawa, Y. and Kodera, Y. (2005) Improvement of Image Quality in Chest MDCT Using Nonlinear Wavelet Shrinkage with Trimmed-Thresholding. Japanese Journal of Radiological Technology, 61, 1599-1608.

11. Lee, Y., Tsai, D.Y. and Suzuki, T. (2008) Contrast Enhancement of Medical Images Using Sigmoid-Type Transfer Curves for Wavelet Coefficient Weighting Adjustment. Medical Imaging and Information Sciences, 25, 48-53.

12. Tsai, D.Y. and Lee, Y. (2003) A Method of Medical Image Enhancement Using Wavelet Coefficient Mapping Functions. Proceedings of IEEE International Conference on Neural Networks and Signal Processing, 2, 1091-1094.

13. Watanabe, H., Tsai, D.Y., Lee, Y., Matsuyama, E. and Kojima, K. (2011) An Integrated Method of Wavelet Coefficient Thresholding for Reducing Radiation Dose While Maintaining Diagnostic Image Quality. Medical Imaging and Information Sciences, 28, 51-56.

14. Chang, S.G., Yu, B. and Vetterli, M. (2000) Adaptive Wavelet Thresholding for Image Denoising and Compression. IEEE Transactions on Image Processing, 9, 1532-1546.

15. Donoho, D.L. and Johnstone, J.M. (1994) Ideal Spatial Adaptation via Wavelet Shrinkage. Biometrika, 81, 425-455.

16. Donoho, D.L. and Johnstone, I.M. (1995) Adapting to Unknown Smoothness via Wavelet Shrinkage. Journal of the American Statistical Association, 90, 1200-1224.

17. Zhang, M. and Gunturk, B.K. (2008) Multiresolution Bilateral Filtering for Image Denoising. IEEE Transactions on Image Processing, 17, 2324-2333.

18. Sendur, L. and Selesnick, I.W. (2002) Bivariate Shrinkage Functions for Wavelet-Based Denoising Exploiting Interscale Dependency. IEEE Transactions on Signal Processing, 50, 2744-2756.

19. Karthikeyan, K. and Chandrasekar, C. (2011) Speckle Noise Reduction of Medical Ultrasound Images Using Bayesshrink Wavelet Threshold. International Journal of Computer Applications, 22, 8-14.

20. Watanabe, H., Tsai, D.Y., Lee, Y., Nakamura, T., Miyazaki, M., Kuramochi, Y. and Kojima, K. (2009) Evaluation of Irreversible Compressed Images in Computed Radiography Using Physical Image Quality Measures. Japanese Journal of Radiological Technology, 65, 1618-1627.

21. Samei, E., Flynn, M.J. and Reimann, D.A. (1998) A Method for Measuring the Presampled MTF of Digital Radiographic Systems Using an Edge Test Device. Medical Physics, 25, 102-113.

22. Flynn, M. and Samei, E. (1999) Experimental Comparison of Noise and Resolution for 2k and 4k Storage Phosphor Radiography Systems. Medical Physics, 26, 1612-1623.

23. Neitzel, U., Gunther-Kohfahl, S., Borasi, G. and Samei, E. (2004) Determination of Detective Quantum Efficiency of a Digital X-Ray Detector: Comparison of Three Evaluations Using a Common Image Data Set. Medical Physics, 31, 2205-2211.

24. Samei, E. and Flynn, M.J. (2002) An Experimental Comparison of Detector Performance for Computed Radiography Systems. Medical Physics, 29, 447-459.

25. Bankman, I.N. (2000) Handbook of Medical Imaging. Academic Press, San Diego, 24-26.

26. Lim, J.S. (1989) Two-Dimensional Signal and Image Processing. Prentice Hall, Englewood Cliffs, 548.

Chapter 8

A Wavelet Transform Method to Detect P and S-Phases in Three Component Seismic Data

Salam Al-Hashmi[1*], Adrian Rawlins[2], Frank Vernon[3]

[1]Earthquake Monitoring Center, Sultan Qaboos University, Muscat, Oman

[2]University of Newcastle upon Tyne, Newcastle upon Tyne, UK

[3]Institute of Geophysics and Planetary Physics, University of California, San Diego, USA

ABSTRACT

The discrete time wavelet transform has been used to develop software that detects seismic P and S-phases. The detection algorithm is based on the enhanced amplitude and polarization information provided by the wavelet transform coefficients of the raw seismic data. The algorithm detects phases, determines arrival times and indicates the seismic event direction from three component seismic data that represents the ground displacement in three orthogonal directions. The essential concept is that strong features of the seismic signal are present in the wavelet coefficients across several scales of time and direction. The P-phase is detected by generating a function using polarization information while S-phase is detected by generating a function based on the transverse to radial amplitude ratio. These functions are shown to be very effective metrics in detecting P and S-phases and for determining their arrival times for low signal-to-noise arrivals. Results are compared with arrival times obtained by a human analyst as well as with a standard STA/LTA algorithm from local and regional earthquakes and found to be consistent.

KEYWORDS

Discrete Time Wavelet Transform; P and S-phases; Automatic Detection; Rectilinearity Function

1. INTRODUCTION

Seismic events such as earthquakes cause a release of energy represented by seismic waves that can be recorded by seismic monitoring stations. Detection of seismic waves and estimation of their arrival times provides information about earthquake location and magnitude. Analysis and detection of seismic waves may be done by visual inspection by a trained analyst or automatic analysis software.

Rapid and accurate detection of seismic waves is of great importance in locating earthquakes [1-10]. Automatic detection techniques are of interest because they can be processed in near real-time; they apply a consistent set of metrics and are repeatable. However, it is often very difficult to determine consistent estimates of P and S waves if they have low signal-to-noise characteristics, particularly if arriving at seismic stations at regional distances.

The objectives of this research are:

1) To design an algorithm for the detection and analysis of the two main types of seismic body waves (P and S-phases).

2) To test the algorithm using seismic events recorded by various stations at local and regional distances, and then compares the results with the results obtained by other methods.

Several methods have been tried recently these that have involved digital signal processing in both time and frequency domains [11-13]. The method adopted here is wavelet transform using an approach initially developed by K. Anant and F. Dowla [11].

The algorithms have been developed in Matlab 6.5™ (http://www.mathworks.com) including signal processing and WaveLab™ (http://www-stat.stanford.edu/~wavelab) toolboxes. WaveLab is a library of Matlab routines for wavelet analysis, wavelet-packet analysis, cosine-packet analysis and matching practice. The algorithms can be run on a laptop with 256 MB RAM or on a mainframe computer.

1.1 Seismic Data Used in Testing

Seismic data from Earthquake Monitoring Center in Oman has been used for testing the algorithms. The ANZA Broadband Seismic Network (http://eqinfo.ucsd.edu) also provided analyst reviewed data and STA/LTA automatic detected data used for comparison with the wavelet detector.

The paper structured as follows: An introduction about wavelet transform is given in Section 2. Section 2 also includes description of the discrete time wavelet transforms, its application on seismic signals to match specific features and how the selection of wavelet type has been done in this study. The multiresolution analysis technique and the perfect reconstruction filter banks are explained too. The developed software and methodology are explained in Section 3. The flow charts of the algorithms are illustrated in Appendix B and a detailed annotation is given in Section 3 too. The results are discussed in Section 4. Comments about the testing results and the comparison results with other methods are presented in Section 5.

2. WAVELET TRANSFORM ANALYSIS

2.1. Introduction

The wavelet transform is a very useful tool for analysing noisy and transient signals and has the ability to represent the signal in both its time and frequency domains. It is a very useful tool in the analysis of non-stationary signals such as seismic signals due to its ability to resolve features at various scales [14].

The wavelet theory arises in 1909 when Haar constructs the first orthonormal system of compactly supported functions called the Haar basis [15].

The term wavelet has been coined in the field of seismology in 1940 by Ricker, N. [15]. The wavelet theory has been applied on seismic signals by Grossmann and Morlet in 1984 [11]. In 1988, Daubechies developed an orthonormal, compactly supported wavelet basis that is smoother than Haar basis [14-16]. Application of wavelet transform on seismic signals has been done by Anant and Dowla [11], Ooninc x, P. J. [12] as well as by Zhang et al. 2003 [13].

The wavelets form a family. The basic form is called the mother wavelet. All the daughter wavelets are derived from this wavelet according the following equation:

$$\Psi_{s,\tau}(t)=\frac{1}{\sqrt{s}}\Psi\left(\frac{t-\tau}{s}\right) \tag{2.1}$$

where, s and τ are the scale and translation of the daughter wavelet. The term $s^{-1/2}$ normalizes the energy for different scales, whereas the other terms define the width and translation of the wavelet.

Digital signal analysis using wavelet transforms begins with the construction of a single parent wavelet. The signal is then decomposed into a series of basis functions of finite length consisting of dilated (stretched) and translated (shifted) versions of this parent wavelet function, i.e., wavelets of different scales and positions in time or space. This process is similar to Fourier analysis, where the parent wavelet is analogous to the sine wave, and the basis functions in Fourier decomposition are sine waves of various amplitude, phase, and frequency variations of the parent sine wave [14, 17].

Scaling a wavelet simply means stretching (or compressing) it. The smaller the scale factor, the more "compressed" the wavelet. The more stretched the wavelet, the longer the portion of the signal with which it is being compared, and thus the coarser the signal features being measured by the wavelet coefficients. Thus, there is a correspondence between wavelet scales and frequency as revealed by wavelet analysis:

- Low scale s ⇒ compressed wavelet ⇒ rapidly changing details ⇒ High frequency ω.

- High scale s ⇒ Stretched wavelet ⇒ Slowly changing, coarse features ⇒ Low frequency ω.

The decomposition advantage is very useful in dealing with signals contain features with various frequency characteristics. Another advantage of the wavelet transform is that its analysis can be chosen based on the application [11].

Figure 1 shows seismic signal at top and wavelet coefficients of six scales where "Daubechies 8" wavelet has been used. The smallest scale is the one that contains the highest frequency while the largest scale is the one that contain the lowest frequency.

2.2. The Discrete Time Wavelet Transform

A discrete type of wavelet transform exists, termed the discrete time wavelet transform (DTWT), where scales and shifts take on discrete values. It allows fast computation of the transform for the digitized signals and gives perfect signal reconstruction. It is a form of multiresolution analysis and is related to perfect reconstruction filter banks [14,18,19].

The wavelet transform can be applied to seismic signals in terms of the type of decomposition as well as in terms of pattern matching [11,12].

In wavelet analysis, we often speak of approximations and details. The approximations are the high-scale, lowfrequency components of the signal while the details are the low-scale, high-frequency components. **Figure 2** shows that the input signal is filtered by a low-pass and high-pass filters in one stage wavelet transform. Then, the signal is down-sampled by a factor of 2 to produce DWT coefficients.

2.2.1. Subband Coding

The DTWT is implemented using the sub-band coding scheme represented in **Figure 3" target="_self"> Figure 3**. The signal is decomposed into 2 frequency bands: low (approximations) and high (details). For the approximation and detail coefficients to represent a discrete-time wavelet decomposition H and G filters must belong to a perfect reconstruction filter bank and required to be regular [14,18,20,21].

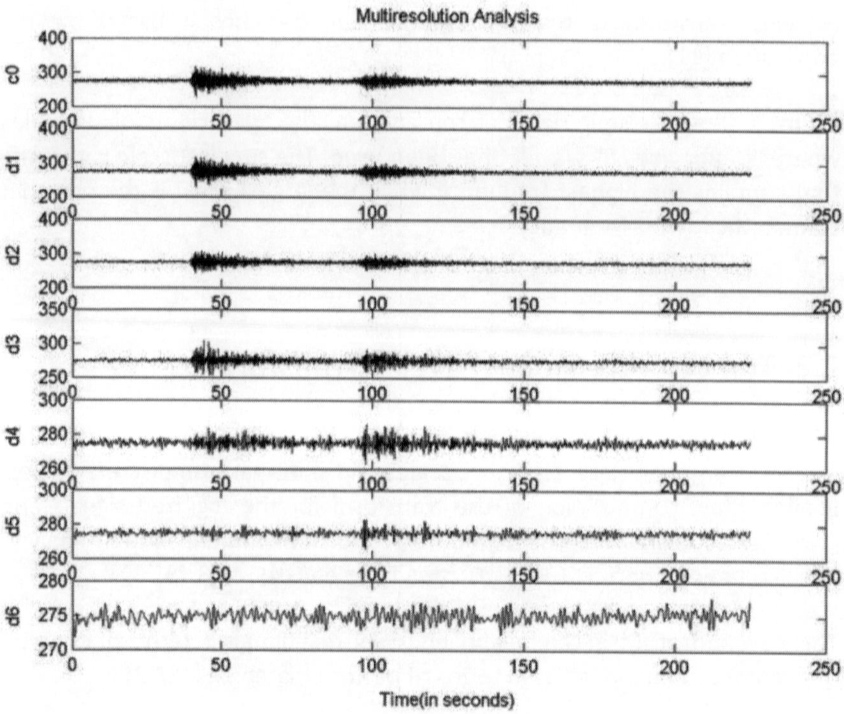

Figure 1. Seismic signal decomposed into several scales.

The wavelet analysis involves filtering and downsampling while the wavelet reconstruction process consists of up-sampling and filtering. Up-sampling is increasing the number of samples by inserting zeros between samples.

The following filters then perform an interpolation by filling in the zero points with the appropriate signal information.

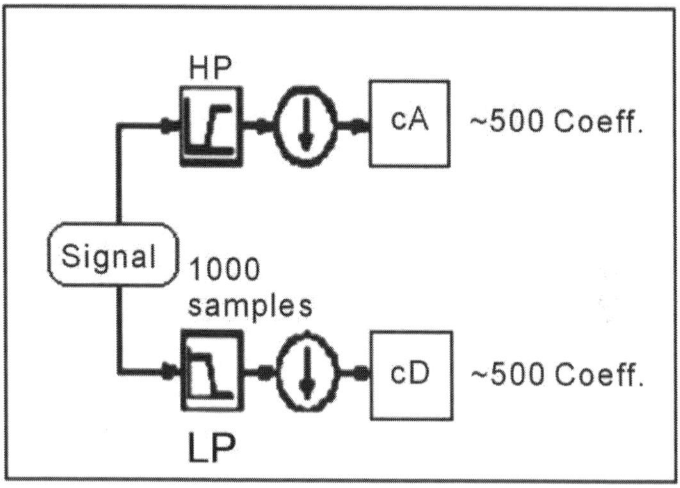

Figure 2. One stage wavelet transforms.

The filtering part of the reconstruction process bears some discussion, because it is the choice of filters that is crucial in achieving perfect reconstruction of the original signal.

Figure 4 is a typical example of such filters magnitude responses.

The down-sampling of the signal components performed during the decomposition phase introduces a distortion called aliasing. It turns out that by carefully choosing filters for the decomposition and reconstruction phases that are closely related, we can "cancel out" the effects of aliasing.

In order for the lowand high-pass decomposition filters (H and G), together with their associated reconstruction filters (L^* and G^*) shown in **Figure 3** to perform such decomposition and reconstruction a system of what is called Quadrature Mirror Filters is used [14,17,19].

2.2.2. Quadrature Mirror Filters

QMF banks are a set of finite impulse response (FIR) filters that enable a signal to be decomposed into subbands and allow reconstruction of the signal from the sub-bands without distortion. It is an example of decimation and interpolation and allows designing two-channel perfect reconstruction filter banks and therefore to generate wavelet bases.

The analysis filters have typically low pass and high pass frequency responses with a cutoff at $\pi/2$ (the "quadrature frequency") as illustrated in **Figure 4**. The philosophy of QMFs is to allow aliasing to be introduced by using overlapping filters for the analysis bank and design the synthesis filters in such a way that any aliasing is exactly cancelled out in the reconstruction process. Filters are also designed so that overall amplitude and phase distortion are minimized or eliminated.

2.3. Wavelet Type Selection

The selection of wavelet type is crucial in the processing. It can depend on seismic arrival shapes. Therefore, the wavelets used in the processing are chosen based on matching the seismic phase's arrival shapes. Thus, the selection will rely on the event as well as on the station that recorded this event. This is because the arrival shapes will vary from event to event and between the different stations as the seismic waves travel through different media and distances for each event and for each monitoring station.

Figure 5 shows some of the wavelets used in the wavelet transform decomposition.

In some cases where P-arrival shape clearly identified, it can be noticed that the wavelet that produce a composite function CT (the function that is used to locate S-arrival time) with the highest magnitude is closest in shape to P-arrival.

Figures 6 and 7 illustrate examples of P-arrival shapes of two different events. In the first example the P-arrival shape seems similar to Daubechies 8 while in the second example the P-arrival shape seems similar to Daubechies 6.

In our analysis the wavelet type selection was verified to be more important with respect to S-phase arrival, while the P-arrival is not affected significantly by the wavelet that is applied. This was also proved by Anant and Dowla [11].

3. SOFTWARE DEVELOPMENT

3.1. Introduction

The software is developed in Matlab 6.5™ including signal processing and Wavelab™ toolboxes. It consists of two algorithms: the P-phase detection algorithm and S-phase detection algorithm described in the sections below. The P-phase detection algorithm can be categorized into three main modules:

- DTWT Processing that includes the following:
    - Multiresolution Analysis
    - Signal Decomposition (For east, north and vertical components).
    - Coefficients Reconstruction (For east, north and vertical components).
- The Composite Rectilinearity Function Construction
- Backazimuth calculation The S-phase detection algorithm can also be categorized into three main modules:
    - Components Rotation
    - DTWT Processing that includes the following:
        - ✓ Multiresolution Analysis
        - ✓ Signal Decomposition (For radial and transverse components).
        - ✓ Coefficients Reconstruction (For radial and transverse components).

- Constructing the Composite Function of Transverse over Radial Amplitude Ratio.

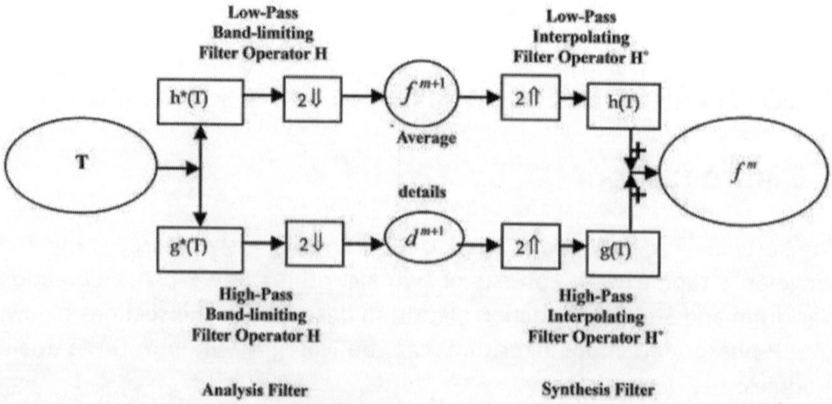

Figure 3. Sub-band coding scheme.

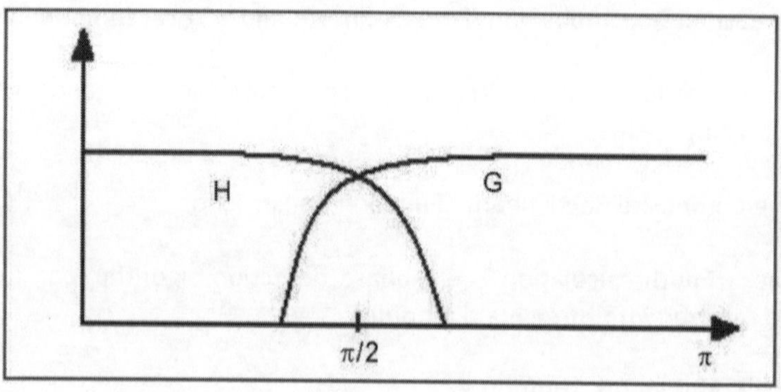

Figure 4. Magnitude responses of analysis filters.

3.2. P-Phase Detection Algorithm

It is known that the P-phase is linearly polarized with respect to the direction of propagation. Therefore, a metric that measures the degree of linear polarization is helpful to detect the P-phase arrival. Such a metric known as the rectilinearity function is defined by Kanasewich [22].

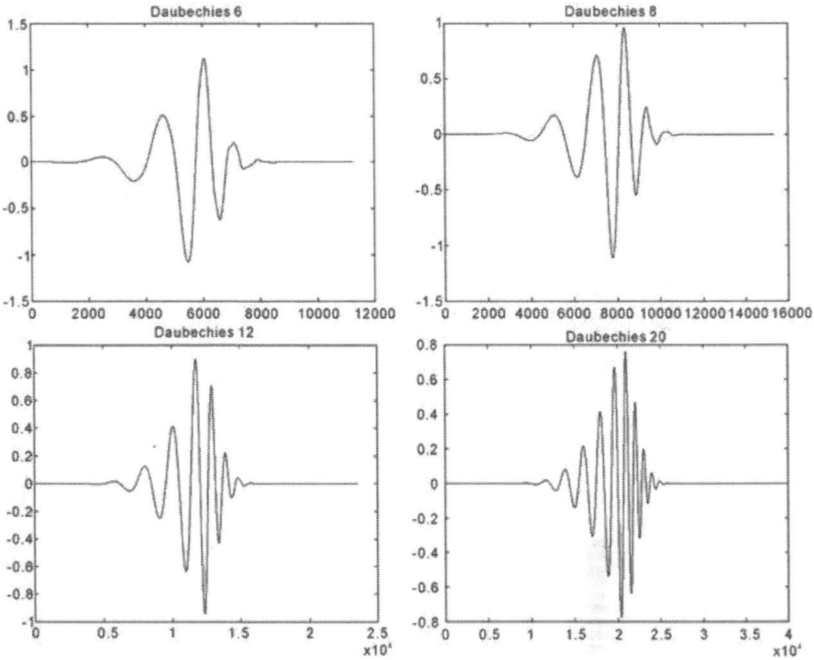

Figure 5. Some of the wavelets used in the research.

Its equation is:

$$\mathbf{F} = \mathbf{1} - (\lambda_2/\lambda_1) \tag{3.1}$$

where λ_1 and λ_2 are the largest and the second largest eigenvalues of the covariance matrix respectively.

If the covariance matrix (Equation 3.2) is diagonalized, an estimate of the rectilinearity of particle motion trajectory over the specified time window can be obtained from the ratio of the principal axis of this matrix [22] i.e. the rectilinearity estimation can be obtained from the ratio of the largest and the second largest eigenvalues of the covariance matrix. The covariance matrix [23] is defined as:

$$\mathbf{M} = \begin{bmatrix} Cov(X,X) & Cov(X,Y) & Cov(X,Z) \\ Cov(Y,X) & Cov(Y,Y) & Cov(Y,Z) \\ Cov(Z,X) & Cov(Z,Y) & Cov(Z,Z) \end{bmatrix} \tag{3.2}$$

whereX: east component wavelet coefficients at scale j;

Y: north component wavelet coefficients at scale j;

Z: vertical component wavelet coefficients at scale j.

While the covariance of X and Y is defined as:

$$Cov(X,Y) = \frac{1}{N}\sum_{i=1}^{N}(X(i)-\mu_X)(Y(i)-\mu_Y) \tag{3.3}$$

where μ_X and μ_Y are the mean values of X and Y respectively.

The direction of polarization may be measured by considering the eigenvector of the largest principal axis [22]. If λ_1 is the largest eigenvalue and λ_2 is the next largest eigenvalue of the covariance matrix, then a function of the form as in (Equation 3.1) would be close to unity when rectilinearity is high ($\lambda_1 \gg \lambda_2$) and close to zero when two principal axes approach one another in magnitude (low rectilinearity).

The direction of polarization can be determined by considering the components of the eigenvector associated with the largest eigenvalue with respect to the coordinate directions X, Y and Z.

The P-phase algorithm is developed as can be described in the following steps:

1) Processing the three component seismic signals using Discrete Time Wavelet Transform (DTWT)

A three component seismic signals represented by east (X), north (Y) and vertical (Z) are processed by the discrete time wavelet transform (DTWT) to calculate the wavelet coefficients wc1, wc2 and wc3 respectively.

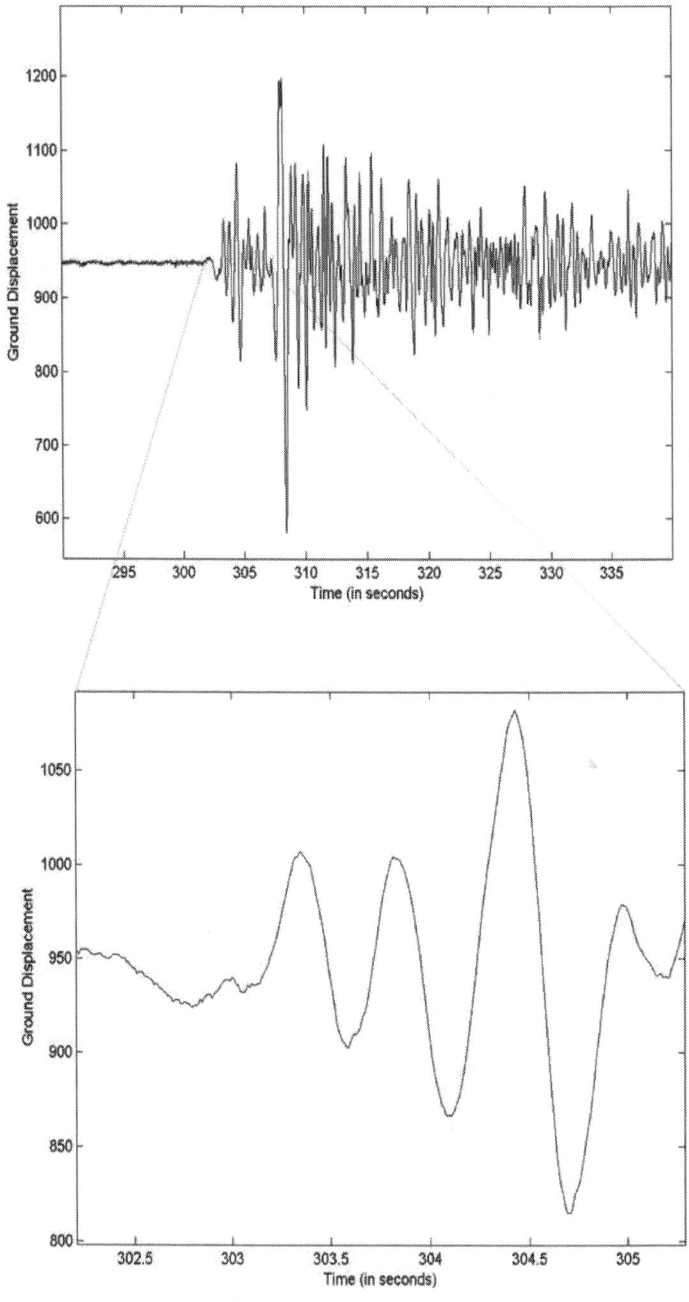

Figure 6. P-phase arrival of an earthquake record and the lower plot is a zoomed view of the arrival.

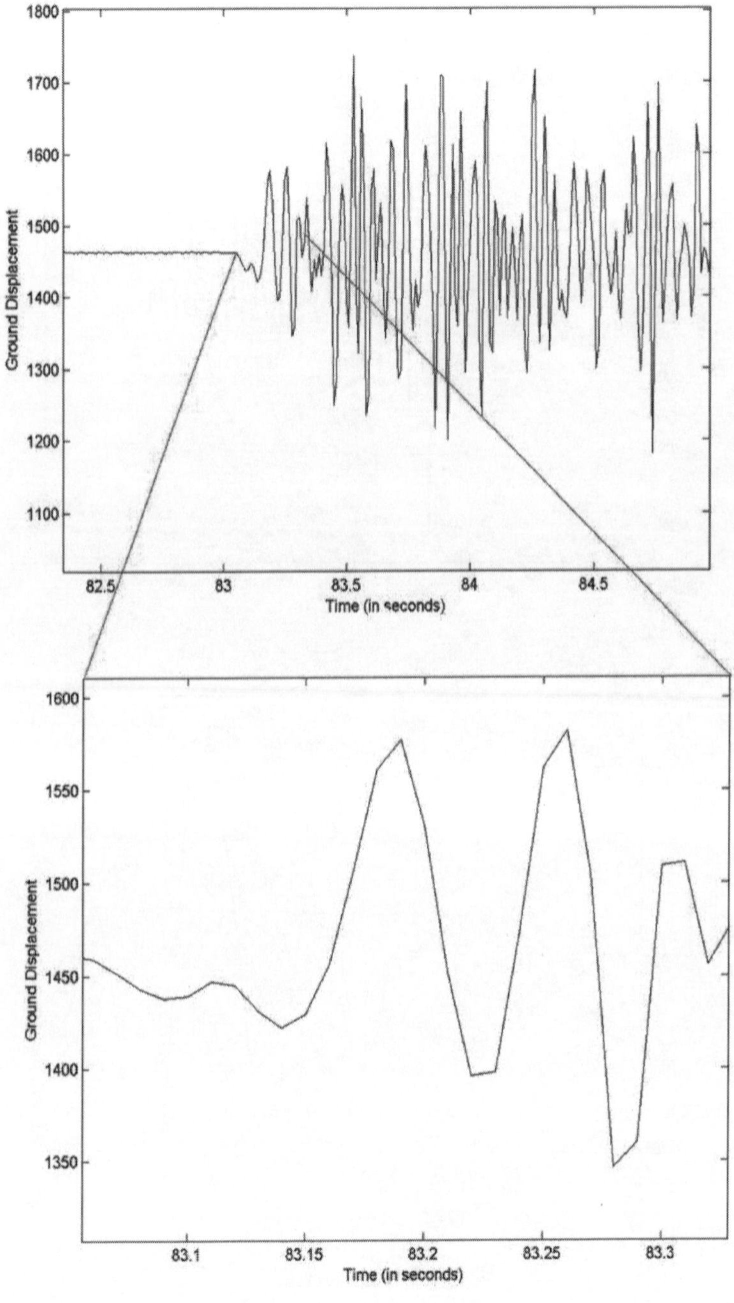

Figure 7. P-phase arrival of an earthquake record and a zoomed view of the arrival.

"Daubechies" wavelets are used in this process because of their compact support basis functions and their shape that coincide with seismic wave arrivals shapes.

Multiresolution Analysis

Each component is decomposed into several scales. So, the results are the wavelet coefficients x_1^j, x_2^j and x_3^j for east, north and vertical components respectively (j is the number of scales).

2) Constructing the rectilinearity function

At each scale, a pointwise moving window is constructed over the three components as shown in**Figure 8**.

The window length is determined using a measure called the varimax norm that is described in Appendix A. At each window, a 3-by-3 covariance matrix is constructed by using Equation (3.2) with x_1^j substituted for X, x_2^j substituted for Y and x_3^j substituted for Z. Then, eigenvalues and their corresponding eigenvectors are calculated.

At each window the rectilinearity function is constructed so that the result is a rectilinearity function (Fj) for each scale. Then, a composite rectilinearity function (Cf) is constructed so that the rectilinearity function of each scale contributes in this function. This composite function is constructed by (Equation 3.4).

$$Cf = \prod_j Fj \tag{3.4}$$

where j is the scale number.

The location where this function gets its maximum value is taken as the P-arrival time.

3) Calculating the back azimuth angle

Back-azimuth is the angle measured from north to the direction from which the energy arrives at a given station [24,25]. It is used to determine the longitudinal and transverse directions for an incoming ray at a prescribed station. For an incident P wave the backazimuth is calculated by constructing the rectilinearity function and finding the eigenvector associated with the maximum eigenvalue for the detected P-wave. This eigenvector represents the direction of linear polarization that specifies the back azimuth angle. This calculation can be conducted at each wavelet scale as well as on the original signals. But, since the original three component signals generally contain noise, the calculation is carried out at the third and higher scales while the first two scales that contain high frequency noise are excluded. It is found that the polarization is conserved along different scales i.e. the same backazimuth angle is obtained in the different scales. So, one value is considered in the processing.

Nwin

Figure 8. Illustration of windowing used to calculate covariance matrices.

3.3. S-Phase Detection Algorithm

The S-phase is a shear wave with particle motion in the transverse direction [22]. Accordingly, it has higher amplitude in the transverse component relative to its amplitude in the radial or vertical component. So, in order to locate the S-phase arrival, the amplitude ratio of the transverse to radial components of the earthquake record is analyzed.

The S-phase algorithm is developed and can be described as follows:

1) Rotating the east and north components

The backazimuth angle (θ) calculated in the previous part is used to rotate the east and north components to radial and transverse components respectively using the following equation:

$$\begin{bmatrix} dr \\ dt \end{bmatrix} = \begin{bmatrix} \sin\theta & \cos\theta \\ -\cos\theta & \sin\theta \end{bmatrix} \begin{bmatrix} X \\ Y \end{bmatrix} \tag{3.5}$$

where dr and dt are the radial and transverse component respectively.

2) Processing the radial and transverse components by the discrete time wavelet transform (DTWT)

The radial and transverse components processed by the DTWT. So, the result is several scales for each component. Referring to the algorithm the result is x^j and y^j that represents the different scales for radial and transverse components respectively (j is the number of scales).

3) Constructing the amplitude ratio

At each scale, a transverse over radial amplitude ratio is calculated as follows:

$$Atr^j = \frac{envt^j}{\left(envt^j + envr^j\right)} \tag{3.6}$$

where $envt^j$ and $envr^j$ are the envelope functions of transverse and radial components respectively.

The envelope function is used to avoid divide-by-zero problems. It is defined as

$$env(x) = \sqrt{x^2 + h^2} \tag{3.7}$$

where h is the Hilbert transform of x.

All scales are combined to construct a second composite function (CT) to be used to locate the S-arrival.

$$\mathbf{CT} = \prod_j Atr^j \tag{3.8}$$

The point after the P arrival time that has a value that is at least one half the maximum of CT is chosen for the S arrival time. The peak of CT itself is not used because it represents the time when S-wave attains its highest amplitude which is few seconds after its arrival time [11].

Several wavelets are used in the wavelet transform decomposition because the choice of wavelet that used in the processing is very important in how well CT locates the S arrival time.

4. RESULTS

4.1. Introduction

The algorithm has been tested by a real seismic data represented by several events recorded on various three components stations. The tested events apart from event 17 are distributed as shown in **Figure 9**. The Figure illustrates the tested events in orange circles as they located according to the stations in dark blue and black triangles for short period and broadband stations respectively. The purpose of this testing is to investigate if the project objectives listed in section 1 are achieved. To measure the performance of the algorithm, the algorithm results are compared with manual inspection results as well as with another automatic algorithm results (STA/ LTA method).

The performance of the system is measured by two ways: the first way is testing the ability of the software to detect P and S waves. However, the second way is comparing the results with another system results to see how accurate the system is.

4.2. P-Phase Detection Algorithm Results

The rectilinearity function proved to be an effective metric in locating the P-arrival time. **Figure 10**shows the rectilinearity functions of six different scales obtained by testing the algorithm with event 21. It can be observed that the rectilinearity function is approximately equal to 1 when the waveform is linearly polarized and equal to zero when there is no linear polarization. This can be seen clearly in the first four scales.

Figure 11 shows the testing result of an earthquake (event 24). It illustrates the composite rectilinearity function Cf and how it is used to locate the P-arrival time.

Figure 12 shows the tested result of another earthquake (event 1) and the composite rectilinearity function Cf. It shows that the peak of this function is used to determine the P-arrival time. It can be observed from the figure that the position at this function is a maximum is where P-phase arrives.

Figure 13 illustrates the direction of Arabian Sea earthquake (event 4) as the result of testing the algorithm by three component data recorded by ABT station (one of the southern stations). Referring to **Figure 9** it can be observed that ABT station is located in Southern Oman, which means that the algorithm indicates the event direction accurately as the event occurred in Arabian Sea.

In order to test the performance of the algorithm, it is tested by an event of magnitude 1.35 that occurs in California-Mexico border region (event 17). The purpose of this testing is to compare the algorithm results with the results obtained by another automatic detector and manual inspection that has been done by an analyst from the Broadband Array data collection center at the University of California, San Diego. The comparison between the analyst result and the algorithm result is illustrated in **Figure 14**. It can be noticed that the results are consistent with each other.

The three components signals illustrated in **Figure 14** are recorded by a station with noisy local conditions (BVDA2 of ANZA network). That's why; the STA/LTA (short term average/long term average) method couldn't give a result for P-arrival time for the data recorded by this station. However, the P-phase detection algorithm gives a good result though of this noise condition.

The algorithm result is compared with the analyst result as well as with STA/LTA results. The comparison shows that the three results are consistent with each other as illustrated in **Figure 15**.

To enhance the algorithm results, they compared with the results of STA/LTA and analyst results of all the events listed in **Table 1**. The time difference in the arrival pick of the algorithm and that of the analyst is shown in **Figure 16**. It can be observed that the mean magnitude of algorithm error is low (0.1952 seconds) as compared with STA/LTA mean magnitude error (0.3982 seconds). This proves that the performance of the algorithm is efficient.

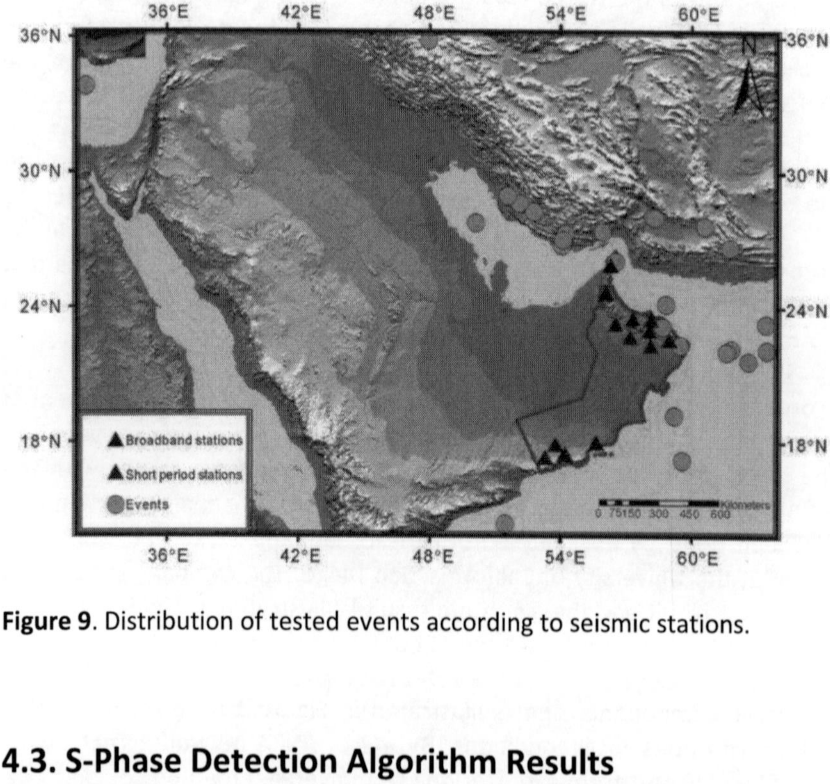

Figure 9. Distribution of tested events according to seismic stations.

4.3. S-Phase Detection Algorithm Results

As stated earlier in Section 3, various wavelets are used in the wavelet transform decomposition. The wavelet is chosen to match the pattern of S-phase to provide the best match in the wavelet scales and thus give the composite function CT with the greatest dynamic range. Therefore, the composite function with the highest amplitude peak is used to locate the S-arrival.

The S-phase detection algorithm is tested by the same events used to test the P-phase detection algorithm. Its testing results of event 17 are compared with the analyst results as shown in **Figure 17** while STA/LTA method failed to give any result for S-arrival.

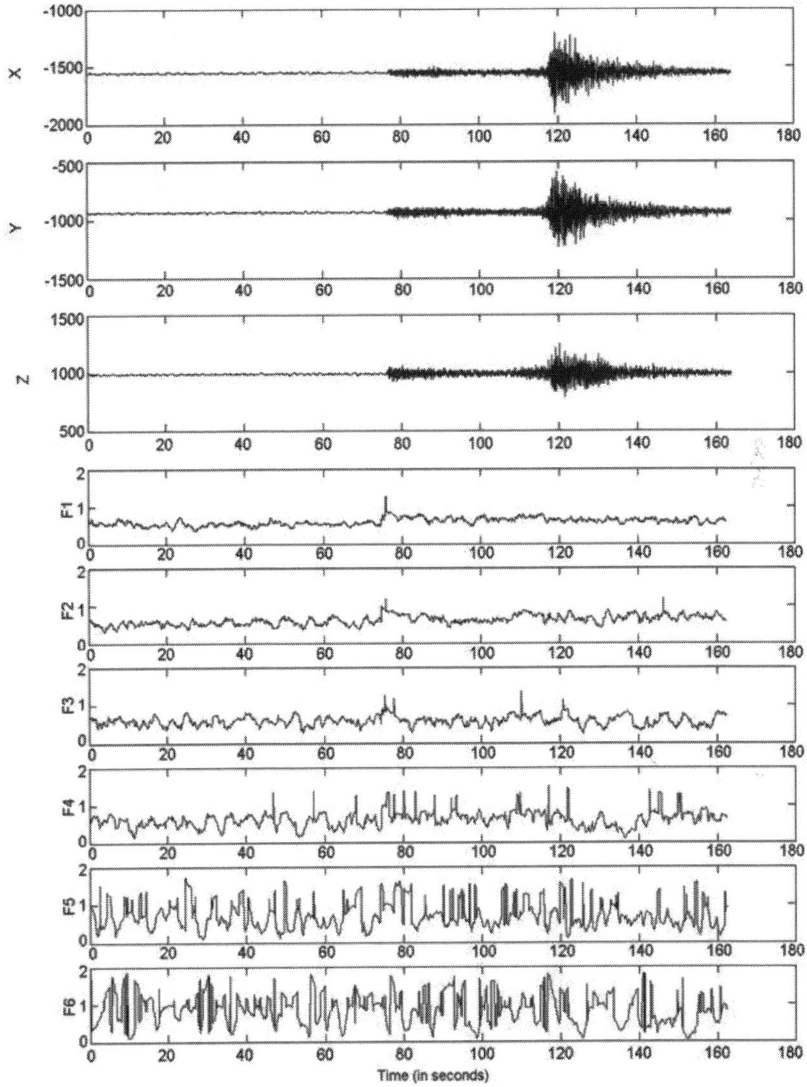

Figure 10. Three component waveforms of event 21 and its related rectilinearity functions of six different scales.

Figures 18 and 19 show the results of testing the algorithm by two different events (event 4 and event 18 respectively). They illustrate the radial and transverse components and the composite function that constructed to be used to detect the S-arrival time. They show that the composite function of transverse to radial amplitude ratio accurately succeeded to determine the S-phase arrival as can be noticed from the figures that the peak of this function indicates the position where S-phase attains its highest magnitude while the onset time of S-phase is few seconds before this peak.

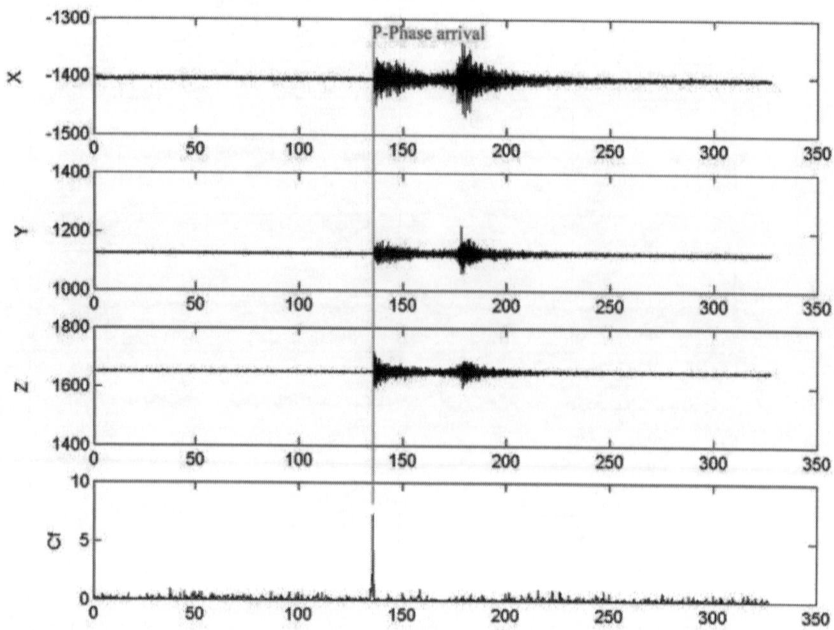

Figure 11. Three component data (east, north and vertical) of event 24 recorded by WBK station (one of the northern stations) with epicentral distance of 386.4 km and backazimuth of 106.9 degrees and the composite rectilinearity function Cf.

As mentioned earlier in Section 4.3, several wavelets are independently used in the wavelet decomposition and the resulting CT of the highest dynamic range has been chosen to determine the S-arrival time. The results show that "Daubechies 8" wavelet is a good choice for most of the tested events. While the other wavelets such as "Daubechies 12" and "Daubechies 20" give good results for some events. This wavelet selection has been done according to the expected arrival shapes as described in details in Section 2.

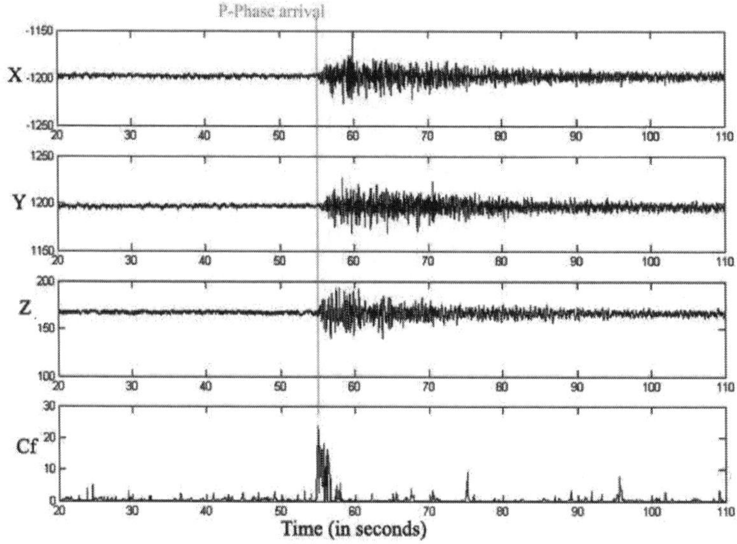

Figure 12. Three component data (east, north and vertical components) of event 1 recorded by JMD station with epicentral distance of 545.898 km and back azimuth of 96.03 degrees and the composite function Cf.

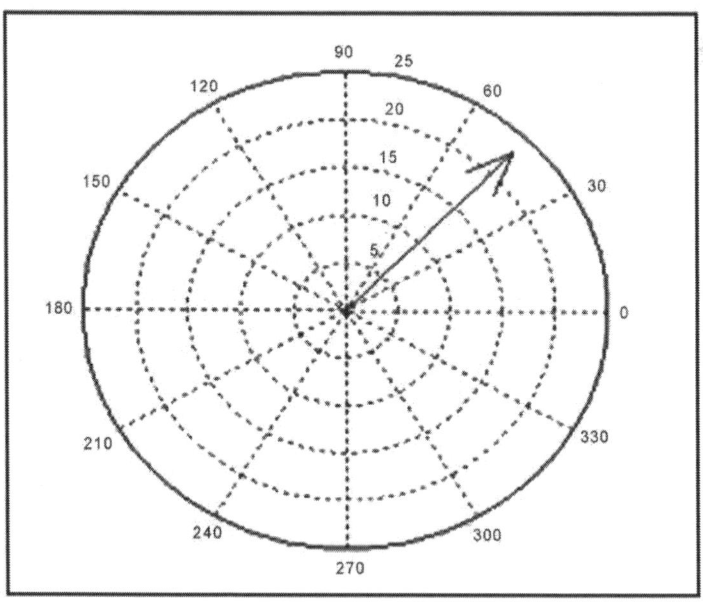

Figure 13. The direction of event 4 from ABT station.

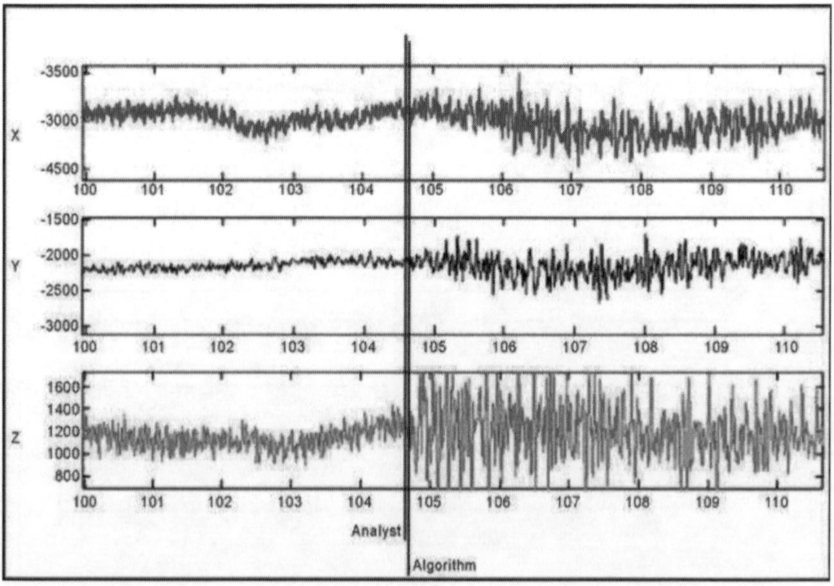

Figure 14. The three component signals of event 17 zoomed around P-arrival time to compare the analyst result with the algorithm result.

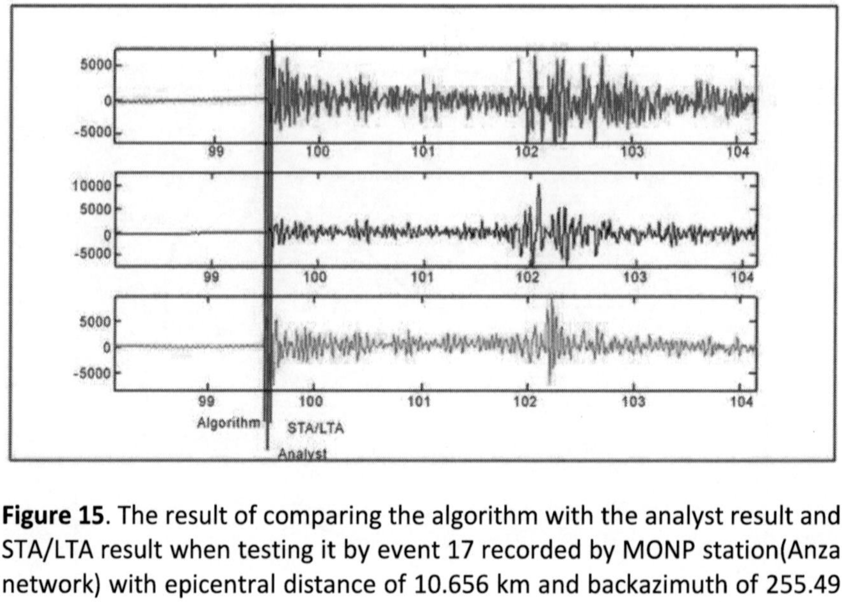

Figure 15. The result of comparing the algorithm with the analyst result and STA/LTA result when testing it by event 17 recorded by MONP station(Anza network) with epicentral distance of 10.656 km and backazimuth of 255.49 degrees.

Table 4. List of tested events.

Event No.	Date of occurrence	Origin time	Latitude	Longitude	Depth	Location
1	01/12/2002	03:43:37	22.11	63.41	30	Arabian Sea
2	02/12/2002	05:00:28	33.8	32.22	60	Eastern Mediterranean
3	03/12/2002	21:42:48	28.26	52.77	50	Southern Iran
4	03/12/2002	23:08:47	17.21	59.56	50	Arabian Sea
5	09/05/2003	00:03:38	26.7817	54.047	50	Southern Iran
6	09/05/2003	00:47:03	27.2357	54.1686	20	Southern Iran
7	09/05/2003	05:54:43	27.4024	55.9109	20	Southern Iran
8	09/05/2003	10:34:07	27.2866	54.2007	35.4	Southern Iran
9	09/05/2003	12:01:30	27.0192	54.0923	30	Southern Iran
10	09/05/2003	16:42:17	28.0641	58.1985	50	Southern Iran
11	09/05/2003	19:17:03	28.6498	52.124	20	Southern Iran
12	10/05/2003	06:39:31	27.6593	60.5815	20	Southern Iran
13	10/05/2003	15:47:46	27.7768	50.1192	30	Persian Gulf
14	11/05/2003	10:17:28	28.9479	51.5855	20	Southern Iran
15	11/05/2003	10:56:14	24.1477	58.8162	36.7	Gulf of Oman
16	11/05/2003	18:01:53	19.1998	59.1707	20	Arabian Sea
17	26/07/2003	05:56:37	15.4463	116.3117	32.92	California-Mexico Border Region
18	24/06/2004	21:56:05	28.0082	58.2898	50	Southern Iran
19	24/07/2004	00:00:01	23.1694	58.662	30	Eastern Arabian Peninsula
20	27/07/2004	05:10:44	23.3008	63.3888	30	Near Coast of Pakistan
21	05/08/2004	08:38:24	14.3831	51.4685	20	Eastern Gulf of Aden
22	29/08/2004	00:43:18	22.1723	61.7808	30	Arabian Sea
23	30/08/2004	08:19:58	26.0658	56.5246	31.8	Southern Iran
24	13/12/2004	01:36:04	21.6386	62.568	35	Arabian Sea
25	13/03/2005	03:31:27	26.6853	61.7142	30	Southern Iran
26	23/03/2005	19:49:33	22.3243	59.4517	13.95	Eastern Arabian Peninsula
27	24/03/2005	00:37:12	22.0547	61.5609	38.6	Arabian Sea
28	03/05/2005	07:20:43	35.8229	47.9509	30	Western Iran

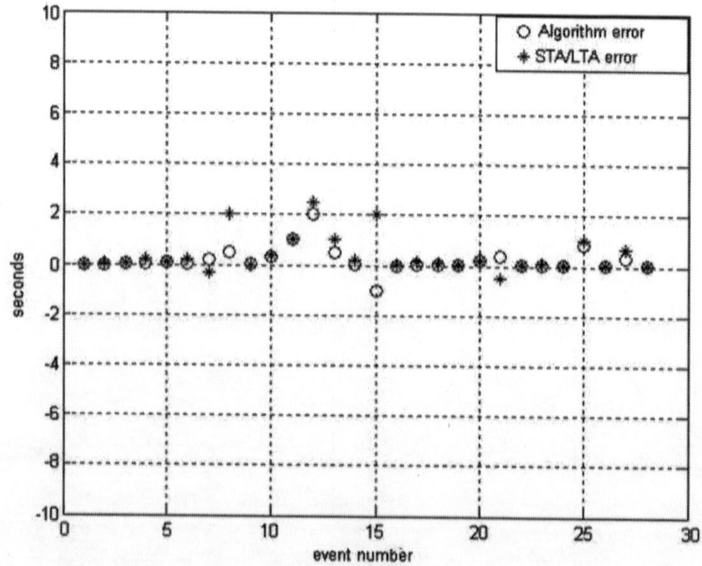

Figure 16. Residual plot of P-Phase arrival.

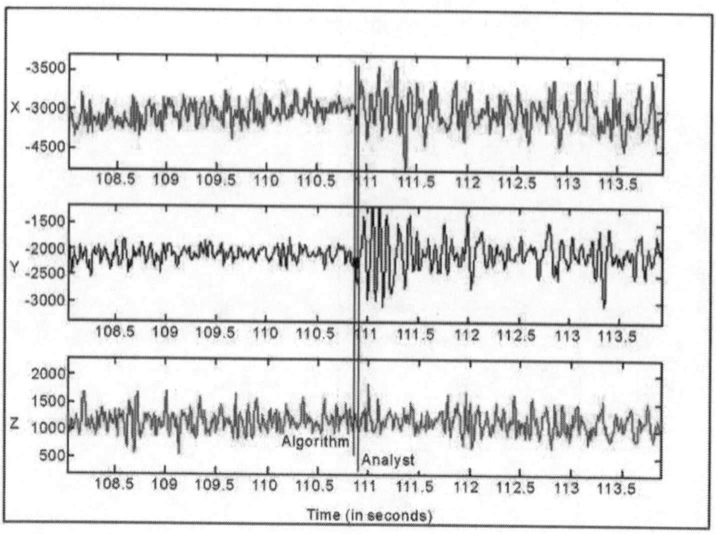

Figure 17. The three component signals of event 17 recorded by BVDA2 station with epicentral distance of 45.732 km and backazimuth of 353.73 degrees, zoomed around S-arrival time to compare the analyst result with the algorithm result.

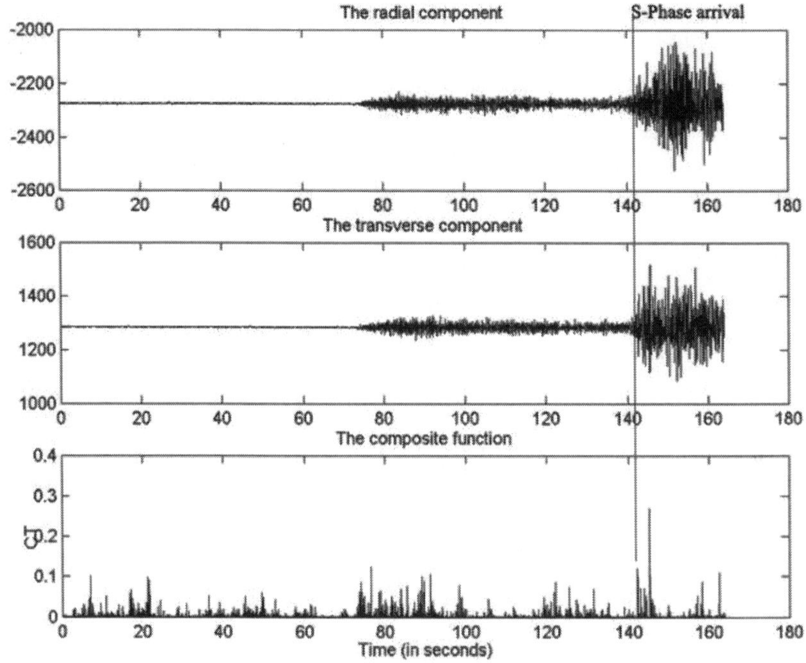

Figure 18. The radial and transverse components of event 4 recorded by ABT station with epicentral distance of 664.224 km and backazimuth of 43.9 degrees, and the composite function of the transverse to radial amplitude ratio.

5. CONCLUSIONS

Since the main objective of the project is to develop an algorithm that automatically detects particular classes of seismic wave, the software has been developed to detect P and S-phases in three component seismic data.

The algorithm has been tested by a real seismic data represented by multiple local and regional events recorded on various three components stations where it successfully detected P and S-phases.

Two important concepts have been presented in this research. The first concept is decomposing the signal into several resolutions or scales; important features in the seismic signals can be identified. Strong features in a signal can be maintained in several scales while weaker features will present in fewer scales or just one scale. The second concept is the wavelet type selection that depends on matching the wavelet to the arrival shapes.

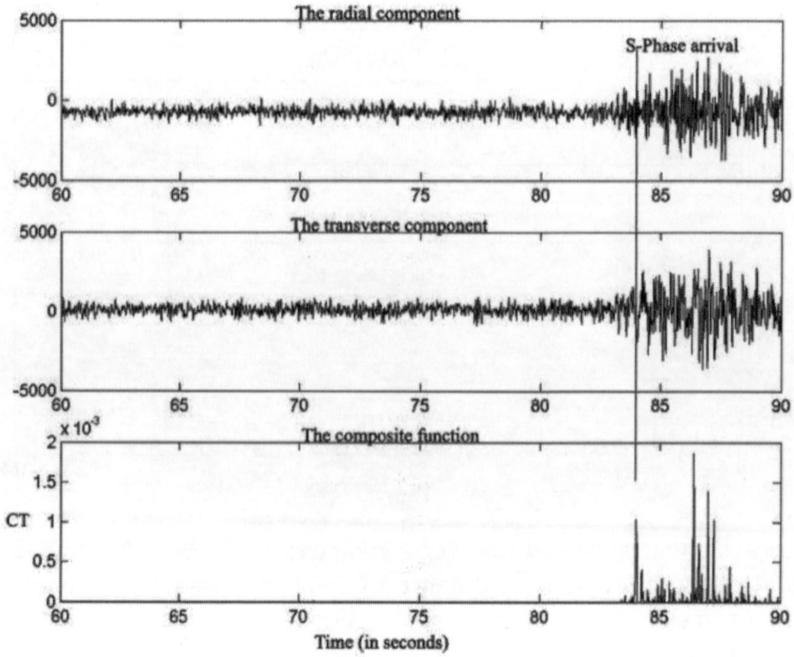

Figure 19. The radial and transverse components of event 18 recorded by BAN station (one of the broadband stations in northern Oman) with epicentral distance of 274.5 and backazimuth of 40.8 degrees and the composite function of transverse to radial amplitude ratio.

The results of the software have been compared with a human analyst results as well as with the results obtained by software that uses STA/LTA (short term average/long term average) method. From the comparison results it can be concluded that the composite rectilinearity function and the composite function of transverse to radial amplitude ratio accurately determined P and S arrivals respectively. The results are consistent with the manual inspection results that done by an analyst as well as with other software based method in current professional use. The results showed that the STA/LTA failed to give any detection for S-phase of the tested events recorded by stations with low signal to noise ratio conditions. In addition, the STA/LTA will fail to give a result for P-phase in low signal-to-noise conditions. However, the P and S-phase wavelet detection algorithms provided good results for P and S-arrivals and the results are consistent with the human analyst results.

To sum up, the developed wavelet algorithm can accurately determine the P and S arrivals in spite of the low signal-to-noise ratio. Furthermore, the software succeeded in determining back azimuths towards the earthquake sources.

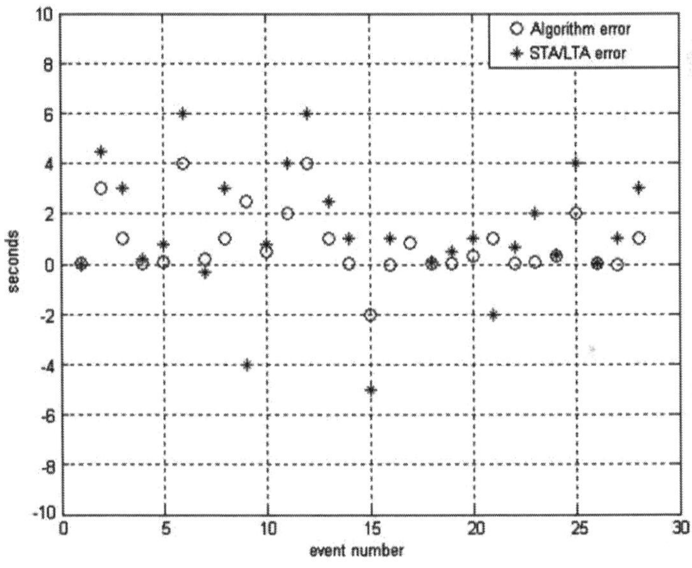

Figure 20 shows the residual plot of S-Phase arrival that illustrates the error between the algorithm and the analyst results (in circles) and the error between STA/ LTA results and the analyst results (in asterisks). It can be noticed that the mean magnitude error of the algorithm error is low (0.8197 seconds). In addition, it can be observed that there is no S-pick from STA/LTA for event 17 while the algorithm has good results and consistent with the analyst results.

The developed software can be used for a daily routine analysis in the seismological laboratories. In order to achieve this it should be tested by a huge amount of data set so that the parameters such as wavelet type that used in the DTWT analysis and the window size that used to construct the covariance matrices can be fixed.

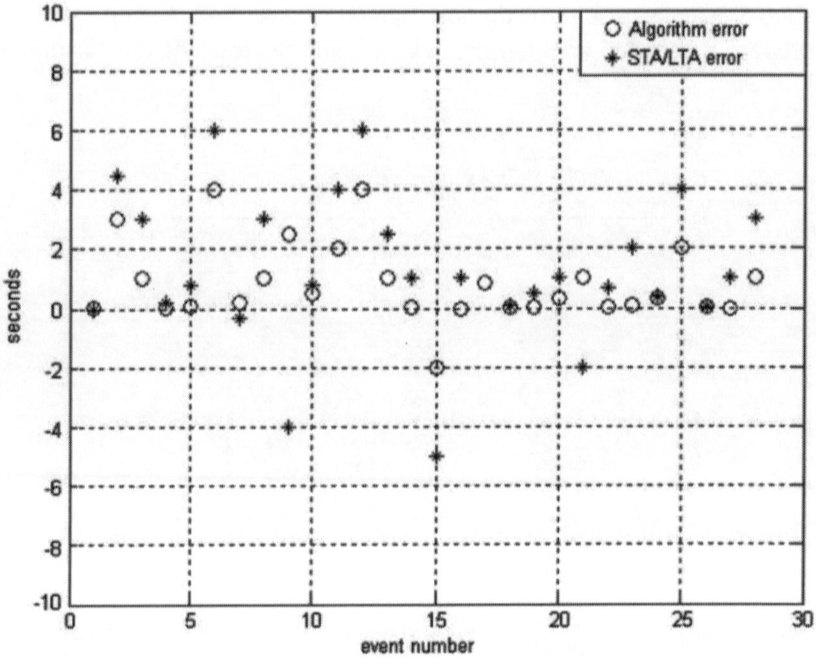

Figure 20. Residual plot of S-phase arrival.

ACKNOWLEDGEMENTS

Sincere thanks to Sultan Qaboos University for sponsoring this study. The first author would like to thank his colleagues at the earthquake monitoring center in Oman for sending the seismic data that are used in testing the software. In addition, many thanks to the staff at the Broadband Array data collection center, UCSD (http://eqinfo.ucsd.edu), for providing the seismic data that are used for results comparison. Partial support for this research was provided by NEHRP/USGS grants 03HQPA-0001 and 01HQAG0021. Furthermore, thanks to the University of Newcastle upon Tyne for providing computer facilities.

REFERENCES

1. H. Dai and C. MacBeth, "Application of Back-Propagation Neural Networks to Identification of Seismic Arrival Types," Physics of the Earth and Planetary Interiors, Vol. 101, No. 3-4, 1997, pp. 177-188.

2. H. Dai and C. MacBeth, "The Application of Back-Propagation Neural Network to Automatic Picking Seismic Arrivals from Single-Component Recordings," Journal of Geophysical Research, Vol. 102, No. B7, 1997, pp. 15105- 15113.

3. R. Allen, "Automatic Phase Pickers: Their Present Use and Future Prospects," Bulletin of the Seismological Society of America, Vol. 72, No. 68, 1982, pp. S225-S242.

4. M. Baer and U. Kradolfer, "An Automatic Phase Picker for Local and Teleseismic Events," Bulletin of the Seismological Society of America, Vol. 77, No. 4, 1987, pp. 1437-1445.

5. A. Cichowicz, "An Automatic S-Phase Picker," Bulletin of the Seismological Society of America, Vol. 83, No. 1, 1993, pp. 180-189.

6. P. S. Earle and P. M. Shearer, "Characterization of Global Seismograms Using an Automatic-Picking Algorithm," Bulletin of the Seismological Society of America, Vol. 84, No. 2, 1994, pp. 366-376.

7. D. Patane and F. Ferrari, "ASDP: A PC-Based Program Using a Multi-Algorithm Approach for Automatic Detection and Location of Local Earthquakes," Physics of the Earth and Planetary Interiors, Vol. 113, No. 1-4, 1999, pp. 57-74.

8. R. Sleeman and T. Van Eck, "Robust Automatic P-Phase Picking: An On-Line Implementation in the Analysis of Broadband Seismogram Recordings," Physics of the Earth & Planetary Interiors, Vol. 113, No. 1-4, 1999, pp. 265-275.

9. P.-J. Chung, M. L. Jost and J. F. Böhme, "Estimation of Seismic-Wave Parameters and Signal Detection Using Maximum Likelihood Methods," Computers and Geosciences, Vol. 27, No. 2, 2001, pp. 147-156.

10. M. Tarvainen, "Automatic Seismogram Analysis: Statistical Phase Picking and Locating Methods Using OneStation Three-Component

Data," Bulletin of the Seismological Society of America, Vol. 82, No. 2, 1992, pp. 860- 869.

11. K. S. Anant and F. U. Dowla, "Wavelet Transform Methods for Phase Identification in Three-Component Seismograms," Bulletin of the Seismological Society of America, Vol. 87, No. 6, 1997, pp. 1598-1612.

12. P. J. Oonincx, "A Wavelet Method for Detecting SWaves in Seismic Data," Computational Geosciences, Vol. 3, No. 2, 1999, pp. 111-134.

13. H. Zhang, C. Thurber and C. Rowe, "Automatic P-Wave Arrival Detection and Picking with Multiscale Wavelet Analysis for Single-Component Recordings," Bulletin of the Seismological Society of America, Vol. 93, No. 5, 2003, pp. 1904-1912.

14. I. Daubechies, "Ten Lectures on Wavelets," The Society of Industrial and Applied Mathematics, 1992.

15. E. J. Stollnitz, T. D. DeRose and D. H. Salesin, "Wavelets for Computer Graphics, Theory and Applications," Morgan Kaufmann, Burlington, 1996.

16. G. Strang and T. Q. Nguyen, "Wavelets and Filter Banks," Revised Edition, Wellesley-Cambridge Press, Wellesley, 1998.

17. S. Mallat, "A Wavelet Tour of Signal Processing," Academic Press, Waltham, 1999.

18. S. Mallat, "A Theory for Multiresolution Signal Decomposition: The Wavelet Representation," IEEE Transactions on Pattern Analysis and Machine Intelligence, Vol. 11, No. 7, 1989, pp. 674-493.

19. B. Porat, "A Course in Digital Signal Processing," John Wiley & Sons, Inc., Canada, 1997.

20. M. Vetterli and C. Herley, "Wavelets and Filter Banks: Theory and Design," IEEE Transactions on Signal Processing, Vol. 40, No. 9. 1992, pp. 2207-2232.

21. M. Vetterli and D. Le Gall, "Perfect Reconstruction FIR Filter Banks: Some Properties and Factorizations," IEEE Transactions on Acoustics, Speech and Signal Processing, Vol. 37, No. 7. 1989, pp. 1057-1071.

22. E. R. Kanasewich, "Time Sequence Analysis in Geophysics," University of Alberta Press, Edmonton, 1981.

23. A. Jurkevics, "Polarization Analysis of Three-Component Array Data," Bulletin of the Seismological Society of America, Vol. 78, No. 5, 1988, pp. 1725-1743.

24. Stein, S. and M. Wysession, "An Introduction to Seismology, Earthquakes and Earth Structure," Blackwell Publishing Ltd., Hoboken, 2003.

25. K. Aki and P. G. Richards, "Quantitative Seismology," University Science Books, Sausalito, 2002.

26. R. Wiggins, "Minimum Entropy Deconvolution," Geoexploration, Vol. 16, No. 1-2, 1978, pp. 21-35.

Chapter 9

Wavelets in Electrochemical Noise Analysis

Peter Planinšič[1] and Aljana Petek[2]

[1] University of Maribor, Faculty of Electrical Engineering and Computer Science, Slovenia

[2] University of Maribor, Faculty of Chemistry and Chemical Engineering, Slovenia

1. INTRODUCTION

Electrochemical noise (EN) is the term used to describe the spontaneous fluctuations of current or potential, which are generated during the corrosion processes. It has been investigated extensively since the 1968, and data has shown the use of EN measurements offers valuable sources of information about complex electrochemical reactions such as those in corrosion systems (Gabrielli et al., 1985; Bertocci & Huet, 1995).

Many methods can be used to analyze the data, such as the variance, standard deviation and root mean square in the time domain. Alternatively, the signal has been transformed from the time domain to the frequency domain using fast Fourier transform or the maximum entropy method, giving the power spectrum density (PSD). The technique of wavelet analysis may be used instead, where a set of wavelets of varying amplitude, duration, and location be constructed such that reproduces the signal of interest.

Wavelets based methods are modern mathematical tools for multiscale time frequency analysis and characterization of in general nonstationary EN signals. This work presents the short overview to usability and possibilities of wavelet transformation in comparison with classic analysis.

2. CORROSION PROCESSES AND ELECTROCHEMICAL NOISE

Corrosion can be defined as the deterioration of materials due to its interaction with its environment and is appearing in various forms: as localized corrosion and as general (uniform) corrosion. Localized corrosion results in the accelerated loss of material at discrete sites in a passive materials surface, leading to the perforation or other failure. Since the majority of the surface is unaffected the difficulty with localized corrosion is its detection and prediction. Uniform corrosion results from the sites that are distributed randomly over the surface regarding both the space and time. Uniform corrosion damage is manifested in dissolving and the progressive thinning of a metal. To prevent corrosion or to predict the outcome of a corrosion situation the knowledge of mechanism of various form of corrosion is fundamental. Thermodynamic principles can be applied to determine which processes can occur and how strong the tendency is for changes to take place. Kinetic laws then describe the rates of the reactions.

Corrosion of metals in aqueous environments is electrochemical in nature. It occurs when two or more electrochemical reactions take place on a metal surface, producing dissolved species of metal or solid corrosion products and thus lowering the energy of the system. The corrosion process has been written as two separate reactions occurring at two distinct sites on the same surface: the anode (metal dissolution site) and the cathode (site of the accompanying reduction reaction). At corrosion of iron in an acid solution containing dissolved oxygen, iron is oxidized to ferrous ion which passes into solution in the anode region

$$Fe \rightarrow Fe^{2+} + 2e^- \tag{1}$$

and H^+ ion is reduced to hydrogen or O_2 is reduced to water in the cathode region

$$2\,H^+ + 2e^- \rightarrow H_2 \tag{2}$$

$$O_2 + 4\,H^+ + 4e^- \rightarrow 2\,H_2O \tag{3}$$

The overall corrosion reaction is the sum of the anodic and cathodic partial reactions:

sum of the reactions (1) and (2)

$$Fe + 2\,H^+ \rightarrow Fe^{2+} + H_2 \tag{4}$$

and sum of the reactions (1) and (3)

$$2\,Fe + O_2 + 4\,H^+ + 4e^- \rightarrow 2\,Fe^{2+} + 2\,H_2O \tag{5}$$

Electrochemical corrosion processes can be investigated by observation of charge flows between the electrolyte and the corroding metal. The measure of the rate at which reducible or oxidizable species can gain or lose electrons is the current density, j i.e. the charge flux trough a metal/electrolyte interface. The potential dependence of charge transfer rate is known as the Butler-Volmer equation:

$$j = j_0 \left\{ e^{(1-\alpha) f \eta} - e^{-\alpha f \eta} \right\} \tag{6}$$

Where j0 is exchange current density, α is transfer coefficient, η is overpotential, i.e. measure of how far the reaction is from equilibrium, and f is defined by:

$$f = \frac{nF}{RT} \tag{7}$$

where *F* is Faraday constant, *R* is the gas constant and *T* is the absolute temperature. If the Butler-Volmer equation is used to express the current densities in corrosion processes, the anodic reaction is the metal dissolution and the cathodic reaction is the accompanying reduction of H^+

or O_2. The equilibrium is achieved (the overpotential is zero) at the corrosion potential where no measurable current j flows: the anodic current density of metal dissolution must be equal to the cathodic current density and is equal to the corrosion current density, j0 . The overpotential is said to be positive if it is such as to produce a positive current, i.e. if it drives the anodic oxidation reaction and suppresses the reduction reaction.

In measurements procedure a potentiostat is often used and the electrochemical cell with tree electrodes: working electrode (WE) represents the interface of interest, the reference electrode (RE) acts as standard for potential measurements and the counter electrode acts as electron sink or source for reactions that occur on the surface of WE. The potential of the WE is controlled with respect to the RE at a constant value, and the current density j under those conditions is determined. If the mean current is compensated or subtracted from the data and only random fluctuations are remained then electrochemical current noise is obtained. Electrochemical current noise thus can be measured as the random fluctuation in current of WE that is held at fixed potential or as the galvanic coupling current between two nominally identical working electrodes. Electrochemical potential noise is measured as the random fluctuation in potential of a WE with respect to a RE or as the fluctuation in potential difference between two nominally identical working electrodes.

In our investigations the current noise was monitored by a low-noise battery-operated potentiostat Jaissle IMP88 PC-R at a sampling rate of 10 points per second. The cell assembly was put in a Faraday cage. An example of measured time series for two processes (pitting as signal I_0 and general corrosion as signal I_2) is presented in Fig. 1. As is seen, the signal has a relatively smooth appearance for general corrosion and occasional sharp increases and decreases in the amplitude of current noise data occur for localized corrosion.

3. STOCHASTIC PROCESSES AND 1/*F* NOISE

EN-signals are generated from corrosion processes which are stochastic in their nature. A short theoretical overview of stochastic processes follows (Schroeder, 1991; Flandrin, 1992; Gao et al., 2007).

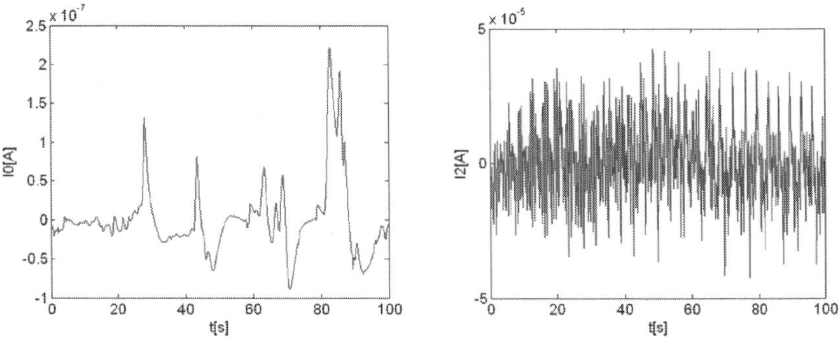

Figure 1. Current noise signal corresponding to: a) X5CrNiMo17-13 stainless steel in borate buffer solution + 0.1 mol/dm^3 NaCl at the passive potential (I_0); b) Low Carbon Steel in 0.1mol/dm^3 H_2SO_4 at the corrosion potential (I_2) (Planinšič & Petek, 2008).

For understanding the stochastic processes it is essential to understand the concept of probability which is associated with random events. Often random events are presented by numbers, called a random variable. Let us denote a random variable by *X*, whose value depends on the outcome of random experiment *ω*. The probability P(Xx) is denoted by $F_x(x)$ which is called the cumulative distribution function (CDF). When exist, it is usually more convenient to use its derivative $f_x(x)$, called the probability density function (PDF):fx(x)=dFx(x)/dx. Some commonly used distributions are normal or Gaussian distributions, exponential and related distributions, binomial and related distributions and heavy tilled distributions. An important class of measures or parameters associated with CDF and PDF for a random variable is expected or mean value.

The mean or average of *X* can be obtained by relation:

$$E[X] = \overline{X} = \int_{-\infty}^{\infty} x dF_X(x) = \int_{-\infty}^{\infty} x f_x(x) \cdot dx \tag{8}$$

The pth order moment of *X* around zero is defined as:

$$E\left[X^p\right] = \int_{-\infty}^{\infty} x^p \cdot dF_X(x) = \int_{-\infty}^{\infty} x^p \cdot f_X(x) \cdot dx \tag{9}$$

The second moment represents the power of a random variable *X*.

The pth order central moment of *X* around mean value is defined as:

$$E\left[\left(x-\overline{X}\right)^{p}\right]=\int_{-\infty}^{\infty}\left(x-\overline{X}\right)^{p}\cdot f_{X}(x)\cdot dx \qquad (10)$$

Variance of *X* is denoted as $Var(X)=\sigma X^{2}$ and is the second order ($p=2$) central moment. The square rot of variance σX is called the standard deviation.

For a a given sample space *S*, a set of events *E* and a probability measure *P*, one can define a stochastic process as follows: For each sample point $\omega \in S$, we assign a time function $X(t,\omega)$. The stochastic process consists of the family of these functions. For each allowable parameter *t*, $X(t, \omega)$ is a random variable. For a fixed ω, $X(t, \omega)$ is a function of a time *t*; it is one realization of the stochastic process. There are many examples of stochastic processes. The well known examples are Markov processes and 1/*f* processes. We will focus our attention to 1/*f* processes. Stochastic process is also called a random process and for simplicity $X(t,\omega)$ is denoted as $X(t)$.

The activity of complex systems can usually be characterized by appearance of $1/f\beta$ noise, a form of temporal fluctuations that has power-law power spectral density property over a wide range of frequencies. β is the power spectral exponent. A convenient framework for studying $1/f\beta$ stochastic process is the self affine stochastic process $X=\{X(t),t\geq 0\}$, which is defined by

$$X(\lambda\cdot t)\overset{d}{=}\lambda^{H}\cdot X(t), t\geq 0 \quad \lambda>0, \; 0<H<1 \qquad (11)$$

where =d denotes equality in distribution, because of using the concept of statistical self similarity of time series. *H* is the Hurst parameter which is the measure of self-similarity. It can be derived, that a mean of such process is:

$$E\left[x(t)\right]=\frac{E\left[X(\lambda t)\right]}{\lambda^{H}} \qquad (12)$$

the variance

$$Var\left[X(t)\right]=\frac{Var\left[X(\lambda t)\right]}{\lambda^{2H}} \qquad (13)$$

and autocorrelation

$$R_{xx}(t,s) = \frac{R_{xx}(\lambda t, \lambda s)}{\lambda^{2H}} \tag{14}$$

By proving that λ−HX(λt) and X(t) have the same power spectral density, one can also prove that the power spectrum density of irregular, self similar processes has 1/*f* property:

$$S(\omega) \approx \frac{\sigma^2}{|\omega|^\beta} \qquad S(f) \approx \frac{\sigma^2}{|2\pi f|^\beta} = \frac{K \cdot \sigma^2}{f^\beta} \tag{15}$$

where *ω* is radial frequency, σ2σ2is the variance, *K* constant value, and *β* the spectral exponent, which defines the slope of power spectral density over several decades.

An example of self-affine stochastic process is fractional Brownian motion process (fBm). It is nonstationary zero mean Gaussian process denoted as BH(t) , characterized by scalar parameter *H*(Hurst parameter). The nonstrationary characteristic of fBm is evident from its covariance function structure:

$$E[B_H(t) \cdot B_H(s)] = \frac{\sigma^2}{2}\left(|t|^{2H} + |s|^{2H} + |t-s|^{2H}\right) \tag{16}$$

where *E* is the expectation operator. From this covariance function follows that the variance is of the type:

$$Var(B_H(t) = E\left[B_H(t)^2\right] = \sigma^2 |t|^{2H} \tag{17}$$

Although fBm process is nonstationary, it has stationary increments, which means that the probability properties of the difference process BH(t)−BH(s) only depend on the lag *t-s*. It is this increment process which is self similar. The slope is in the range ≤β≤3 . The slope *β* is 2 for the classic example Brownian motion. The other example of 1/fβ processes is fractional Gaussian noise (fGn), with −1≤β≤1 . fGn is stationary process. White Gaussian noise has the slopeβ=0 . It can be shown that Brownian motion (β=2β=2) is simply the integral of white noise.It was reported, that *β* is related to the Hurst parameter *H*, which measures statistical self similar properties of signals:

$$\beta = 2 \cdot H + 1 \tag{18}$$

This is the reason for studying 1/fβ noise via self-affine stochastic processes. It was also shown, that for one dimensional signal, *H* is related to fractal dimension *D* by:

$$D = 2 - H \tag{19}$$

D is noninteger parameter in the range 1D2and *H* in the range 0H1. The fractal dimension can also be used for characterizing the complexity of the stochastic signal. Fractals are mathematical sets, which have a high degree of complex geometrical self similarity and can model many kinds of complex time series. The concept of statistical self-similarity and fractals was extended to time series to describe irregular characteristics of signals, from white noise to Brownian motion. The irregularity of a fractal curve or signal can be measured by capacity or fractal dimension *D*, a simplification of Hausdorf dimension, which is easier to calculate numerically. The roughness of such curve depends on *D*. The straight line have dimension 1. The more irregular the curve, the closer is its dimension to 2. There are many definitions and methods for calculation the fractal dimension. We will give a definition on which basis the popular box counting method. Let *S* be a bounded set in ℜn. The minimum number *N*(*s*) of balls of radius *s* is needed to cover *S*;N(s)≈s−DN(s)≈s−D. The fractal dimension is then defined as:

$$D = -\liminf_{s \to 0} \frac{\log N(s)}{\log(s)} \tag{20}$$

4. CLASSICAL STATISTICAL AND FOURIER ANALYSIS METHODS

In the past, the most common EN-analysis methods were statistical and Fourier methods. These methods assume the stationary or quasi stationary nature of processes and signals under consideration.

An early overview of different EN-data analysis methods was made in the work by R. A. Cottis (Cottis, 2001). Follows a little extended theoretical overview of classical methods (Orfanidis, 1996).

4.1. Background Of Statistical And Fourier Methods

By analyzing random processes the statistical parameters as mean value and moments are defined by expectation operators, i.e. by statistical averaging of many realizations of stochastic process. In practice this is many times impossible and there is available only one block or array of *N* time signal samples. The statistical averaging is then replaced by the estimation obtained using sample or time average.

The pth moment of sample $x(n)$; $n=0,..N-1$ is defined as:

$$moment(p,x) = \frac{1}{N}\sum_{n=0}^{N-1} x^p(n) \tag{21}$$

The first moment is mean value $\bar{x}$. The square root of second moment gives the root mean square value x_{rsm}, which measures the amplitude of the signal. The square of x_{rsm} represents the signal power. The pth central moment of sample is defined as:

$$central\_moment(p,x) = \frac{1}{N-1}\sum_{n=0}^{N-1}\left(x(n)-\bar{x}\right)^p \tag{22}$$

The second order central moment is the well known signal variance. The square root of variance is standard deviation, which is usually used for describing the amplitude of noise signals.

Classical spectral analysis bases on Fourier transform. The Fourier Transform of the deterministic continuous time signal $x(t)$ of duration T_1 is defined as :

$$X_{T_1}(j\omega) = \int_{T_1} x(t)\cdot e^{-j\omega\cdot t}dt \tag{23}$$

The Discrete Time Fourier Transform (DTFT) of the deterministic sampled signal with *N* samples is defined as:

$$X_N(\omega) = \sum_{n=0}^{N-1} x(n)\cdot e^{-j\cdot\omega\cdot n\cdot T} \tag{24}$$

where $x(n) = x(nT)$; $n=0,1,\ldots,N-1$, is according to sampling theorem sampled analog signal $x(t)$, n is time index, and T the sampling period. It can be efficient computed by the Discrete Fourier Transform (DFT) and its fast version the Fast Fourier Transform (FFT).

$$X(k) = DFT\{x(n)\} = \sum_{n=0}^{N-1} x(n)\cdot e^{-j\cdot k\cdot n\cdot 2\cdot \pi/N}; k=0,1\ldots,N-1 \tag{25}$$

where n is time index and k is the frequency index. The corresponding frequency resolution is given by:

$$\Delta\omega = \omega_s / N \tag{26}$$

where ω_s is radial sampling frequency. The main shortcoming of classical Fourier transform is the averaging the features across the whole time domain.

EN signals are of stochastic nature; therefore sampled EN signals are random sequences. To obtain smooth spectra an ensemble averaging should be introduced and the spectrum calculated over autocorrelation function. The autocorrelation function of a zero mean random signal is defined as:

$$R_{xx}(k) = E[x(n+k)\cdot x(n)] \tag{27}$$

where E is the averaging or expectation operator. For stationary signals, R_{xx} do not depend on time n, but only on the relative time lag k between sequences $x(n)$ and $x(n+k)$. The power spectrum of the random signal $x(n)$ is defined as the Discrete Time Fourier Transform (DTFT) of its autocorrelation function $R_{xx}(k)$:

$$S_{xx}(\omega) = \sum_{k=-\infty}^{\infty} R_{xx}(k)\cdot e^{-j\cdot\omega\cdot n\cdot T} \tag{28}$$

where ω is the frequency in radians per sec. This power spectrum shows how the power is spread over frequencies and is also called PSD (Power Spectral Density).

EN measurements cannot often be repeated to obtain smoothed spectra by ensemble averaging. One can compute an estimate of expected or true value by so-called sample autocorrelation using time average:

$$\hat{R}_{xx}(k) = \frac{1}{N} \sum_{n=0}^{N-1-k} x(n+k) \cdot x(n) \tag{29}$$

for k=0,1,...N-1. It is known that R^xx(k)is an even function of the lag k. It is also well known that the results are statistical reliable only for small value of lag (5 to 10 percentages).The DTFT of R^xx(k)is S^xx(ω) and is referred to as periodogram spectrum and can be viewed as an estimate of power spectrum:

$$\hat{S}_{xx}(\omega) = \sum_{k=-\infty}^{\infty} \hat{R}_{xx}(k) \cdot e^{-j \cdot \omega \cdot n \cdot T} \tag{30}$$

Using the above equations we can express the periodogram also as:

$$\hat{S}_{xx}(\omega) = \frac{1}{N} \cdot |X_N(\omega)|^2 \tag{31}$$

where XN(ω) is DFTF of N signal samples. It can be efficient computed using FFT. For wide sense stationary random signals the mean of periodogram converges to the true power spectrum Sxx(ω)Sxx(ω) in the limit for large N:

$$S_{xx}(\omega) = \lim_{N\to\infty} E\left[\hat{S}_{xx}(\omega)\right] = \lim_{N\to\infty} E\left[\frac{1}{N} \cdot |X_N(\omega)|^2\right] \tag{32}$$

There are some problems with such classical Fourier spectral analysis method. To achieve high statistical reliability, very long signal sequences should be used. But long signal sequences can no longer be stationary. However, the main shortcoming is the averaging the futures over the whole time domain.

This have lead researchers to find and develop of an advanced signal analysis methods. Recently wavelet based methods for signal analysis found to be useful for nonstacionary signals. Therefore in this overview chapter we will consider wavelet-based methods for EN-signals analysis.

4.2. Overview Of Works Using Classical Methods

In individual systems, the correlations between noise measurements and corrosion processes have been reported by many authors but only some can be mentioned here. The EN data for a passive system (SS 316L/Ringer's solution) and several active systems (mild steel/NaCl, brass/NaCl, Al 6061/NaCl and Al 2024/NaCl) have been analyzed in the frequency domain using power spectral density (PSD) and spectral plots, obtained from the ratio of PSD plots of the potential and current fluctuations. Comparisons of spectral noise spectra with traditional impedance spectra have been made and good agreement has been observed for all systems after trend removal (Lee & Mansfeld, 1998; Mansfeld et al., 2001). Current fluctuation during general corrosion was analyzed upon a simple model, derived on the assumption that elementary fluctuation sources are related to the fluxes of electrons that are transferred from the metal to electron-acceptor ions in solution. The number of successful electron transfers obeyed a Gaussian distribution, from which the corrosion current density and transfer coefficients were determined (Petek et al., 1997; Petek & Doleček, 2001). The time-series noise patterns of the steel in bicarbonate solution (the simulated geological environment) were transformed into frequency domain by fast Fourier transformations, and then their power spectral densities at a frequency were determined to be compared with the corrosion rate (Haruna et al., 2003). Two new indices (S_E and S_G) were derived to evaluate pitting corrosion by dimensional analysis of three parameters of PSD, the slope of high frequency linear region, the critical frequency and the low-frequency plateau level. As shown, the value of S_E can be related to the fluctuation velocity, which can represent the pitting corrosion rate and S_G should contain some information about slow corrosion processes (Shi et al., 2008). PSD had been employed to analyze EN data associated with corrosion behavior of A291D magnesium alloy in alkaline chloride solution. Three corrosion stages, the anodic dissolution process companying with the growth, absorption and desorption of hydrogen bubbles, the development of pitting corrosion, and the inhibition process by protective MgH_2 film could be distinguished. However, the results obtained only from PSD was insufficient for better understanding the corrosion mechanism of alloy during the immersion and the wavelet transform was carried out (Zhang et al., 2007).

4.3. Our Applications Of Classical Methods

EN signal (Fig. 1) is represented as a time series, where one can easily distinguish the fluctuations but not the intensity and frequencies of fluctuations. In the paper (Planinšič & Petek, 2003) we analyzed EN corrosion signals also with some classical methods, which use correlation functions and histograms.Figure 2 shows estimated autocorrelation functions of two corrosion signals I_0 and I_2, respectively.

Noise data were transformed into frequency domain using FFT algorithm and presented as PSD inFigure 3. PSD of current noise data for pitting process exhibited two parts: a low-frequency plateau and high-frequency part, and the roll-off frequency, which is the frequency to separate the two parts of PSD. PSD plot of general corrosion can be characterized by "white noise" which is independent of frequency.

5. WAVELET MULTIRESOLUTION ANALYSIS METHODS

The assumption of stationary behavior of corrosion processes and random signals is not always correct. Corrosion signals are a non stationary in general.

When we are interested on how signal frequency components vary with time, we should use joint time-frequency analysis. For this purpose we can use Window Fourier Transform (WTF), also called Short Time Fourier Transform (STFT) or spectrogram. It is known that the STFT can be considered as the filter bank, consists of Finite Impulse Response Filters (FIR) with equal bandwidth or equal frequency resolution. Therefore it is difficult to meet sharp localization in time and frequency simultaneously. For this reason, this technique is not always appropriate for analyzing natural signals or phenomena, where in the signal exist long duration low frequency components and short high frequency components at the same time. This problem can be elegantly solved using modern multiresolution time frequency analysis methods based on wavelets. It was shown that the Discrete Wavelet Transform can be viewed and realized as multirate filter bank with octave, also called constant Q frequency resolution.

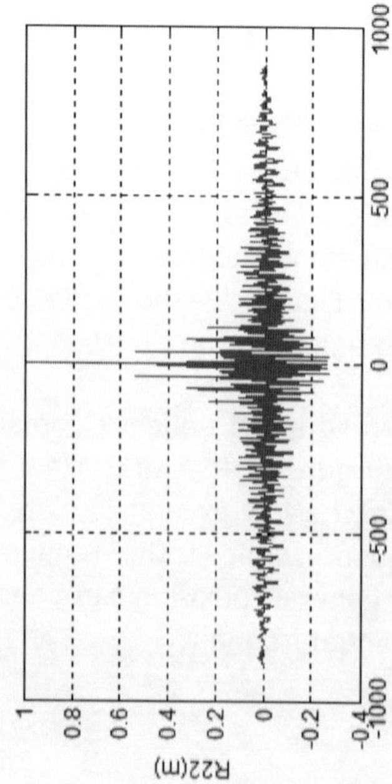

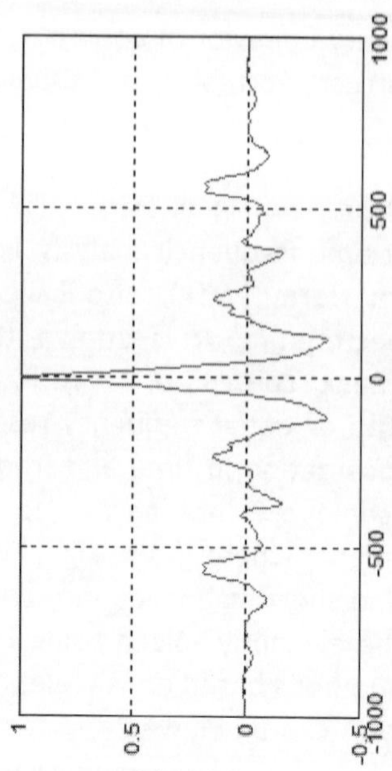

Figure 2. Estimated autocorrelation functions of EN-signals: a) I_0; b) I_2

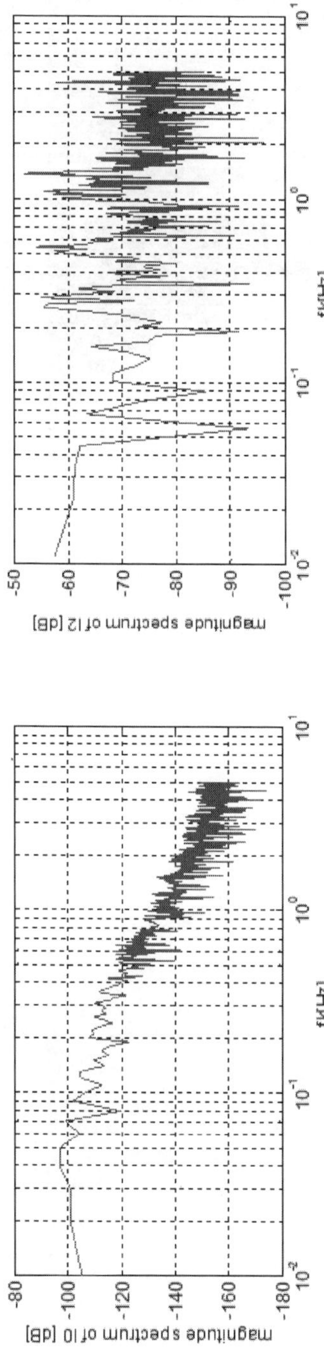

Figure 3. Estimated power spectral density of signals I_0 (left) and I_2 (right)

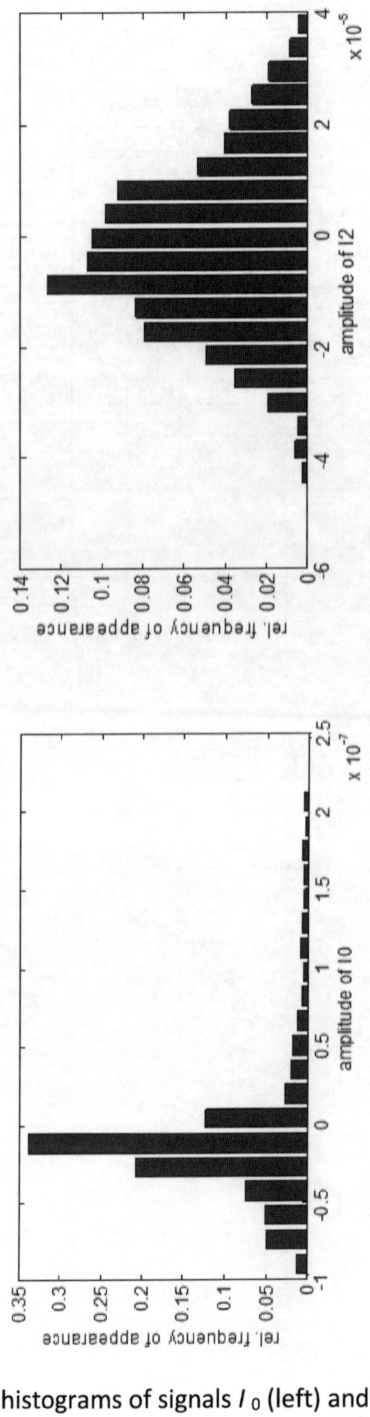

Figure 4. Normalized histograms of signals I_0 (left) and I_2 (right)

Amplitude distribution was studied using normalized histograms. As demonstrated by Figure 4, a current noise amplitude distribution of general corrosion is Gaussian.

A short theoretical overview of wavelet methods follows (Daubechies, 1992; Burrus, 1992;Fladrin,1992, 1993; Radolphe, 1994; Wornell, 1996; Mallat, 1998; Dai et al., 1994; Palawajjhalla et al., 1994).

5.1. Background Of Wavelet Methods

Wavelets are waves which construct basis of signal decomposition in wavelet transforms, similar as trigonometric functions with different frequencies in Fourier Transform. Wavelets are scaled and shifted versions of the so called mother or primary wavelet function ψ(t) . Thus the family of functions is then defined as:

$$\{\psi_{a,b}(t)\}; \quad \psi_{a,b}(t) = |a|^{(-1/2)} \cdot \psi\left(\frac{t-b}{a}\right) \tag{33}$$

where parameters *a* and *b* (a,b∈R;a≠0a,b∈$\Re$;a≠0) are called dilation (scaling) and translation (shifting) parameters, respectively. To be a good analyzing function, the mother wavelets must fulfill some conditions. The first is the so called "admissibility" condition:

$$\int_{-\infty}^{\infty} \frac{|\hat{\psi}(\omega)|^2}{|\omega|} \cdot d\omega < \infty \tag{34}$$

where ψ^(ω) is the Fourier transform ofψ(t)ψ(t). Because the mother wavelet is absolutely inferable functions, we can show that:

$$\hat{\psi}(0) = 0 \Leftrightarrow \int_{-\infty}^{\infty} \psi(t) \cdot dt = 0 \tag{35}$$

Admissibility implies that a wavelet must be an oscillatory decaying function with zero mean. There are also additional other desirable properties for a function to be a useful wavelet, as smoothness, good time and frequency localization, number of vanishing moments. These properties suggest that wavelets are bandpass filters. ψ(t)is the impulse response of filter ψ^(ω).

In the contrast with Fourier analysis where basis functions are trigonometric functions, by wavelet-based analysis different kind of mother wavelet function can be used, appropriate for particular application. There are many types of wavelet transforms according to input signals, time and scaling parameters, used wavelet functions, namely continuous, discrete, bi-orthogonal and semi-orthogonal and orthonormal bases version.

However wavelet transform can be broadly classified into Continuous Wavelet Transform (CWT) and Discrete Wavelet Transform (DWT). CWT of a function f(t)∈L2(R)f(t)∈L2(ℜ)involves the computation of scalar product. Wavelet coefficients are computed as:

$$C_{a,b} = C(a,b) = \int_{-\infty}^{\infty} f(t) \cdot \psi_{a,b}(t) \cdot dt \tag{36}$$

Discrete wavelet transform involves discretization of parameters, *a* and *b*, respectively:

$$a = a_0^{\,m} \qquad b = n \cdot b_0 \cdot a \tag{37}$$

$$\psi_{m.n}(t) = a_0^{-m/2} \psi(a_0^{-m} \cdot t - n \cdot b_0) \tag{38}$$

$$C_{m,n} = \langle f(t), \psi_{m,n}(t) \rangle = \int_{-\infty}^{\infty} f(t) \cdot \psi_{m,n}(t) \cdot dt \tag{39}$$

where Cm,n are called discrete wavelet coefficients. Discrete wavelets ψm.n(t) that satisfy the condition:

$$A \cdot \|f(t)\|^2 \leq \sum_{m,n} \left| \langle f(t), \psi_{m,n}(t) \rangle \right| \leq B \cdot \|f(t)\|^2 \tag{40}$$

are called frames (Daubechies,1992) and form Riesz basis. Discrete wavelets can be further classified into orthogonal, semi-orthogonal or non-orthogonal.

To obtain orthonormal basis, one can chose samples on dyadic grid (base 2):

$$a = 2^m \qquad b = n \cdot 2^m \qquad m \in Z, n \in Z \tag{41}$$

Orthonormal bases and orthonormal wavelet transform, play an important role in theory and practice of multiresolution analysis. The DWT can be further classified into Wavelet Series Transform (WST), when analyzed signal is continuous (*f* (*t*)), and into Discrete Time Wavelet Transform (DTWT), if the signal is time discrete (*f*(*n*)). One possibility of constructing wavelets is using a scaling function φ(t)and multiresolution analysis (Mallat, 1998). Namely, multiresolution algorithm is a natural way of constructing orthogonal wavelets. Multiresolution analysis is decomposition of square integrable functions f(t)∈L2(R) into closes subspaces Vj, where coarser subspace Vj is contained in finer subspace Vj+1Vj+1. The subspaces also satisfy separation condition (∩m∈ZVm={0}∩m∈ZVm={0}) and condition for completeness (∪m∈ZVm∈L2(R).∪m∈ZVm∈L2(ℜ).Additionally, the functions f(t) satisfy the scaling property (f(t)∈Vm⇔f(2t)∈ Vm−1f(t)∈Vm⇔f(2t)∈Vm−1). And, there exist a scaling function in the coarsest spaceφ(t)∈V0φ(t)∈V0, so that the family of functions{φm,n}, φm,n=2−m/2φ(2−mt−n)φm,n=2−m/2φ(2−mt−n), form the so called Riesz basis of subspaceVmVm. Since φ(t)∈V0⊂V1φ(t)∈V0⊂V1 and the φ(2t)φ(2t)is a basis for the subspace V1V1, we can write scaling function as linear combination with the so called two scale difference equation:

$$\phi(t) = \sum_k h(k) \cdot \phi(2t - k) \tag{42}$$

where *h*(*k*) is a finite sequence. It can be shown, that the frequency response of scaling function is a lowpass filter and *h*(*k*) form the lowpass FIR-filter coefficients. Define Wm−1Wm−1 as the orthogonal complement of subspace Vm−1Vm−1 inVmVm, than the direct sum of infinite subspaces VjVj is the whole spaceL2(R)L2(ℜ). The subspace Wm−1Wm−1 contains the detail information needed to go from approximation of function at coarser to finer resolution level *j*. The multiresoluton analysis allows to approximate the given function *f* (*t*) by $f_j(t)$ at each coarser subspace or resolution level. If ψ(t)ψ(t)is a Riesz basis of space W0⊂V1W0⊂V1 , we can also write:

$$\psi(t) = \sum_k g(k) \cdot \psi(2t - k) \tag{43}$$

where the finite sequence $g(k)$ form the highpass (bandpass) filter coefficients, as the frequency response of wavelet is like that of band-pass filter.

The multiresolution analysis form the theoretical basis for fast Discrete Wavelet Transform (fast DWT), using discrete signals *f*(*n*), Zn∈Z , that is sampled version of *f*(*t*). It was introduced by Mallat by the so called pyramidal multiresolution algorithm, where the signal *f*(*n*) is decomposed into *J*decomposition levels. The idea of multiresolution analysis is to write a function as a limit of successive approximations, each of which is a smoothed version.

The sequences at scale *j* can be computed from sequences at scale *j*-1 by the following multirate filter or convolution, followed by subsampling by 2:

$$a_{2^j}(n) = \langle f(t), \varphi_{m,n}(t) \rangle = \sum_k g(k - 2^j \cdot n) \cdot a_{2^{j-1}}$$

$$c_{2^j}(n) = \langle f(t), \psi_{m,n}(t) \rangle = \sum_k h(k - 2^j \cdot n) \cdot a_{2^{j-1}} \quad (44)$$

where c2j(k)=c(j,k)c2j(k)=c(j,k) are wavelet or detailed coefficients and a2j(k)=a(j,k)a2j(k)=a(j,k) are scaling or approximation coefficients. Data sequence a21(k)=a(1,k)a21(k)=a(1,k) at scale *j*=1 represents approximated or smoothed version of the original signal. The sequence c21(k)=c(1,k) c21(k)=c(1,k) at level *j* = 1 represents difference or detail information. The above equations together describe *j* th level analysis filter bank. This calculation is repeated (iterated) up to scale *J* forming the multiresolution pyramidal algorithm; one stage is shown in Figure 5. As was mentioned, the lowpass filter is associated by scaling function and highpass filter by wavelet function. The filters for calculating the synthesis are the same by using orthogonal wavelet transform. Analysis and synthesis filters can have different length, as by using biorthogonal filter banks. However, this algorithm can be used for orthogonal and nonorthogonal wavelets.

For EN-signal originating from corrosion process, wavelet transform decompose it into approximation and detail signal components at different scales and locations. The wavelet transform is therefore convenient tool to analyze the self-similarity of 1/f time series.

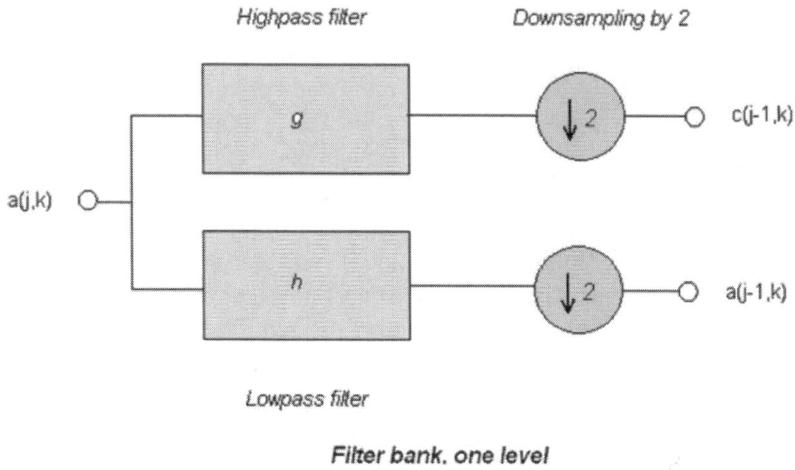

Figure 5. One stage of analysis filter bank

The orthonormal wavelet transform based methods were used for estimating slope β, parameter *H*, and fractal dimension *D* (Akay,1998; Sekine, 2002; Planinšič & Petek, 2008). For orthonormal discrete wavelet decomposition the 1/f property can be replaced by the relation:

$$\sigma_j^2 = \frac{\sigma^2}{\left(2^j\right)^\beta} \tag{45}$$

where σ_j^2 is the variance of detail signal jd2. Then the slope *β* can be calculated from the plot log2σj2 versus level *j*, what can be obtained after short calculation:

$$\log_2 \sigma_j^2 = -\beta \cdot j + \log_2 \sigma^2 \tag{46}$$

5.2. Overview Of Works Using Wavelets

Wavelets have found many applications in different natural scientific disciplines, among them also in chemical engineering (Radolphe et al., 1994; Banjanin et al., 2001). The use of wavelets to study electrochemical noise transients was reported by Aballe (Aballe et al., 1999; 2001). The wavelet analysis of electrochemical noise signals, where the signal was decomposed into wavelet-subbands was used for the characterization of pitting corrosion intensity (Smulko et al., 2002). Wharton et al. demonstrated how the wavelet variance exponent can be used to evaluate corrosion behavior for variety of stainless steels in chloride medium, i.e. be able to discriminate between various corrosion processes covering a wide range of EN signals (Wharton et al., 2003). Wavelet analysis based on the fractional energy contribution of smooth crystals and the lowest frequency detail crystal can provide information on the type and onset of corrosion (general corrosion, metastable pitting, stable pitting) in performed potentiostatic critical pitting temperature test for a superduplex stainless steel (Kim, 2007). In study of the copper anode passivation by electrochemical noise analysis using wavelet transforms it has been found that during active dissolution the electrode surface is dominated by long time scale process and the change of the position of the maximum relative energy from D7 to D8 could be an indication of future passivation (Lafront et al., 2010). It was shown, that electrochemical potential noise analysis of Cu-BTA system using wavelet transformation can be used to achieve the inhibition efficiency (Attarchi et al., 2009).

Some other authors also reported about the fractal nature of corrosion processes and corresponding electrochemical noise signals. The electrochemical potential and current noise originating from the corrosion of carbon steel in distilled water was analyzed using multifractal analysis. The multifractal spectra are found to be qualitatively different for different temporal stages of the corrosion process (Muniandy et al., 2011).

5.3. Our Applications Of Wavelets Methods

Our applications of wavelets transformation or combination with classical methods for the electrochemical current noise analysis were reported for different corrosion processes in several publications (Planinšič & Petek, 2003; 2004; 2007; 2008). For little more detailed insight the short overview of this research is as follows.

Daubechies wavelets "db2" were used to transform the EN signal from Fig. 1. The discrete wavelet transform (DWT) decomposition of signal into on joint time (position) and frequency (scale level) depended amplitudes are presented with color lightness in time-frequency plane in Fig. 6.

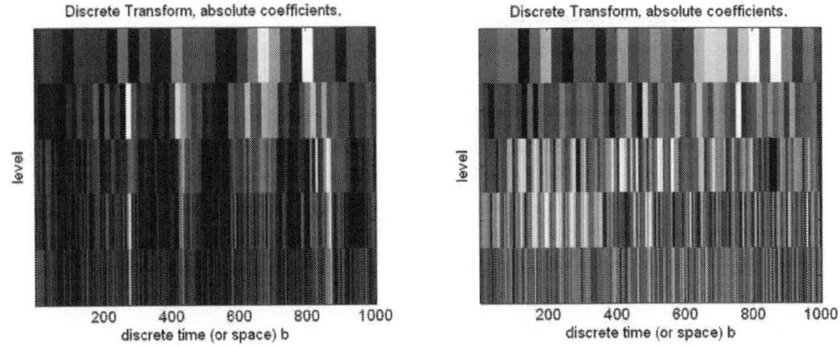

Figure 6. Discrete wavelet transforms (DWT) of signals: a) I_0, b) I_2 (Planinšič & Petek, 2008)

Next, the DWT multiresolution decomposition of processes on 5 levels are shown in Fig. 7. The crystals from D_1 to D_5 are the details of the signal, and A_5 is the approximation of the signal. The frequency range which takes into account each series of detail coefficients, can be computed from relation fs/2/2j where f_s is the sampling frequency and j stands for the corresponding scale.

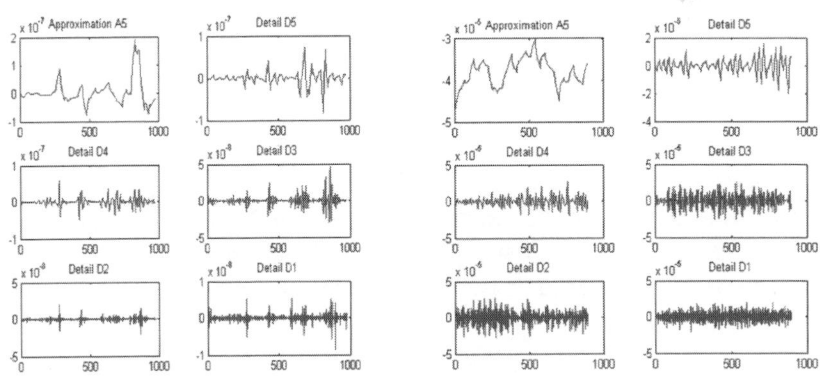

Figure 7. Multiresolution decomposition of discrete signals: a) I_0, b) I_2 on approximation signal A_j and detail signals D_j on five levels (Planinšič & Petek, 2008).

Events with small time constants are taken into account by the fine scale coefficients, details D_1, D_2. The information dealing with larger time constant events is included in details D_4 and D_5. Therefore, these kinds of plot allow the signal to be viewed over the full time range and considering different scales, which contains information about corrosion events occurring at a determined time – scale.

Variances of details were calculated to detect the intensity of particular signal components on level *j*.Fig. 8 shows variances as a function of the decomposition level and also the logarithmic plot of details variances versus level *j* for the slope β determination.

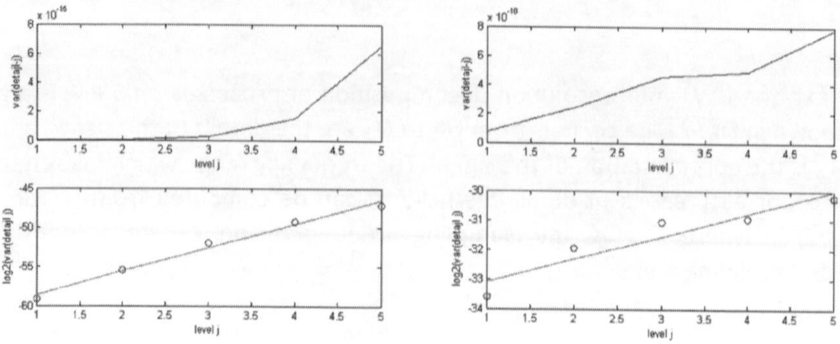

Figure 8. Variances of details plotted as a function of the decomposition level *j* for two processes: a) I_0 (slopeβ = 3.0092) and b) I_2 (slope β = 0.7700) (Planinšič & Petek, 2008).

For the time series I_2 and I_0 were β = 0.7700 and β = 3.0092, respectively. As the value of βincreases, the contributions of high-frequency components in time series are reduced. It is suggested that the events in the relatively higher frequency region may be associated with uniform corrosion. On the other hand, the events in the relatively lower frequency regions are responsible for pitting corrosion.

After short computation the value for Hurst parameter *H*, and fractal dimension *D* can be obtained:

$$D_i = 2,5 - 0,5 \cdot \beta_i \qquad H_i = 0,5 \cdot (\beta_i - 1)\,; \quad i = 1,2. \tag{47}$$

The obtained slope *ß* is the estimated power spectral exponent *ß*. It can be associated with the strength of persistence within a time series. The persistence defines the correlation between adjacent values within time series. If ≤β≤1, than persistence is weak. For the time series with *β* between 2 and 3, the persistence is strong. Fractional Gaussian noise with *β* between -1 and 1, and fractional Brownian motion, with *β* between 1 and 3, are considered as proper representatives of such processes. The first process is stationary and the second is non-stationary.

The obtained results indicating the presence of fractional Brownian motion in pitting corrosion with adjacent values in the time series being strongly correlated and fractional Gaussian noise in general corrosion, with adjacent values in the time series being weak correlated. The Hurst parameter in case of pitting is greater than ½, indicating also the persistence, i.e. a dependence of new values on old values. A summary of the wavelet – based fractal analysis is given in Table 1.

Table 1. The slope *β*, the Hurst parameter *H*, and the fractal dimension *D*, for two corrosion signals (Planinšič & Petek, 2008).

	β_i	H_i	D_i	process	persistence
I_0 – pitting corrosion	3.0092	1.0046	0.9954	non-stationary	strong
I_2 – general corrosion	0.7700	-0.1150	2.1150	stationary	weak

We proposed also a new way for determination of persistence nature of the electrochemical noise on the basis of correlation coefficients between original signal and details $R(I_i,D_j)$, Table 2. The pitting corrosion is positively correlated with long memory effect. Increasing correlation of signal I_2 to the D_2 detail and then decreasing to D_5, indicates weak persistence and short memory effect of general corrosion processes.

Correlation coefficients between successive details $R(D_j,D_{j+1})$ for two decomposed corrosion signals (Table 3) are all zero on the basis of analysis procedure.

Table 2. Correlation coefficients between original signal and details, $R(I_i,D_j)$, for two decomposed corrosion signals (Planinšič & Petek, 2008).

	$R(I_0,D_j)$	$R(I_2,D_j)$
D_1	0.0188	0.4196
D_2	0.0482	0.5303
D_3	0.1128	0.5097
D_4	0.2048	0.3648
D_5	0.3103	0.3138

Table 3. Correlation coefficients between successive details, $R(D_j,D_{j+1})$, for two decomposed corrosion signals (Planinšič & Petek, 2008).

	I_0	I_2
$R(D_1,D_2)$	0.0006	0.0006
$R(D_2,D_3)$	0.0000	0.0004
$R(D_2,D_3)$	0.0001	0.0004
$R(D_4,D_5)$	-0.0012	-0.0120

Additionally, DWT with 3-decomposition levels was made using different kinds of wavelet functions, from Daubechie's fractal-like wavelet "db2" (Massopust, 1994) to smoother wavelet functions, as Daubechie's wavelet function "db5" and symmetrical Coiflet wavelet function "coif5". After decomposition the coding gain (c_g) was calculated from variances of decomposed sub-bands:

$$c_g = \sum_{j=1}^{J}(\sigma_j^2 \cdot \alpha_j) / \prod_{j=1}^{J}(\sigma_j^2)^{\alpha_j} \tag{48}$$

where σj2 are variances of sub-bands and αj are the relative length of sub-band sequences and *J* the number of sub-bands. The coding gain is a measure of spectral flatness. For (uniform) white noise it has the value 1. Also the Shannon's entropy was calculated and can be viewed as a measure of signal complexity. The numerical experimental results are collected in Table 4.

Table 4. Coding gain and Shannon's entropy obtained with different wavelet functions

	c_g, I_2	c_g, white noise	entropy, I_2	entropy, white noise
db2	1.3486	1.0090	4.5384	4.9304
db5	1.3411	1.0076	4.5384	4.9304
coif5	1.3411	1.0116	4.5384	4.9304

The chose of different wavelets did not influence on the obtained c_g and entropy. We found also, that the coding gain and entropy can be used as an additional parameter to distinguish the corrosion processes.

To study the approximation properties of DWT's using different wavelet basis functions, a synthesis (inverse DWT) to different approximation levels was made. We expected better results with "db2" assuming the fractal–like shapes of EN-signals. However, no significant differences were found, a smaller approximation error was obtained even using smoother wavelets, what confirms the approximation theory.

6. CONCLUSION

The most attractive prospective benefit of EN measurement is the ability to obtain information about the type of corrosion that is occurring, but there is much less agreement about the optimum analysis method for obtaining such information (Cottis, 2006). Wavelet transform has been developed over a number of years and only recently has been applied to electrochemical noise analysis. The main advantage of wavelet analysis of EN is the detection of transients which are localized in both the time and frequency domain and shows promise to be discriminatory for the intensity as well as the type of corrosion.

REFERENCES

1. A. Aballe, M. Bethencount, F. J. Botana, M. Markos, 1999Using wavelets transform in the analysis of electrochemical noise data, Electrochimica Acta, 44 26September 1999), 4805 4816 0013-4686

2. A. Aballe, M. Bethencount, F. J. Botana, M. Markos, J. M. Sanchez-Amaya, 2001Use of wavelets to study electrochemical noise transients, Electrochimica Acta, 46 15April 2001), 2353 2361 0013-4686

3. M. Akay, E. J. H. Mulder, 1998Effects of Maternal Alcohol Intake on Fractal Properties in Human Fetal Breathing Dynamics, IEEE Transaction on Biomedical Engineering, 45 9September 1998), 1097 1103 0018-9294

4. M. Attarchi, M. S. Roshan, S. Norouzi, S. K. Sadrnezhad, 2009Electrochemical potential noise analysis of Cu-BTA system using wavelet transformation, Journal of Electroanalytical Chemistry, 633 1 240 245 1572-6657

5. B. Banjanin, B. Gergič, P. Planinšič, Ž. Čučej, 2001Entropy-threshold method for best basis selection, Image and Vision Computing Elsevier, 19 7May 2001), 477 484 0262-8856

6. C. S. Burrus, 1992 Introduction to wavelets and wavelet transforms, A primer. Prentice Hall, 0-13489-600-9Saddle River, New Jersey, USA

7. R. A. Cottis, 2001Interpretation of Electrochemical Noise Data, Corrosion, 57 3 265 285 0010-9312

8. R. A. Cottis, 2006Sources of Electrochemical Noise in Corroding Systems, Russian Journal of Electrochemistry, 42 5 497 505 1023-1935

9. X. Dai, B. Joseph, R. L. Motard, 1994Introduction to Wavelet Transform and Time-Frequency Analysis, In: Wavelet Applications in Chemical Engineering, Motard, R. L. &Joseph, B., (Eds.), 1 32Cluwer Academic Publisher, 0-79239-461-5Massachusetts, USA.

10. I. Daubechies, 1992 Ten Lectures on Wavelets, Siam, 0-89871-274-2Pennsylvania, USA

11. M. Farge, 1993 Wavelets, Fractals, and Fourier Transforms, J. C. Hunt, J. C. Vassilicos (Eds.), Clarendon Press, 0198536470xford, USA

12. P. Flandrin, 1992Wavelet analysis and synthesis of fractional Brownian motion, IEEE Trans. Information theory, 38 2Part 2, Mar. 1992, 904 909 0018-9448

13. P. Flandrin, 1993Fractional Brownian Motion and Wavelets, In: Wavelets, Fractals, and Fourier Transforms, Farge, M., Hunt, J. C. R. & Vassilicos, J. C., (Eds.), 109 142Clarendon Press, 019853647Oxford, USA

14. Y. Gao, Y. Cao, W. Tung, W. , J. Hu, 2007 Multiscale analysis of complex time series (Integration of chaos and random fractal Theory, and Beyond), John Wiley & Sons, 978-0-47165-470-4New Jersey, USA

15. T. Haruna, Y. Morikawa, S. Fujimoto, T. Shibata, 2003Electrochemical noise analysis for estimation of corrosion rate of carbon steel in bicarbonate solution, Corrosion Science, 45 9 2093 2104 0001-0938X

16. J. J. Kim, 2007Wavelet analysis of potentiostatic electrochemical noise, Materials Letters, 61 18 4000 4002 0016-7577X

17. A. M. Lafront, F. Safizadeh, E. Ghali, G. Houlach, 2010Study of the copper anode Passivation by electrochemical noise analysis using spectral and wavelet transforms, Electrochimica Acta, 55 22 2505 2512 0013-4686

18. C. C. Lee, F. Mansfeld, 1998Analysis of electrochemical noise data for a passive system in the frequency domain, Corrosion Science, 40 6 959 962 0001-0938X

19. S. Mallat, 1998 A wavelet tour of signal processing, Academic Press, 1998. 012466606San Diego, California, USA

20. F. Mansfeld, Z. Sun, C. H. Hsu, 2001Electrochemical noise analysis (ENA) for active and passive systems in chloride media, Electrochimica Acta, 46 24-25 3651 3664 0013-4686

21. P. R. Massopust, 1994 Fractal Functions, Fractal Surfaces, and Wavelets, Academic Press, 0-12478-840-8Diego, California, USA

22. Mathworks Inc. 2007Matlab® 7.5.0 (R2007b), 2007

23. S. V. Muniandy, W. X. Chew, C. S. Kan, 2011Multifractal modeling of electrochemical noise of corrosion of carbon steel, Corrosion Science, 53 1 188 200 0001-0938X

24. S. J. Orfanidis, 1996 Introduction to Signal Processing, Prentice Hall, 0-13209-172-0Saddle River, New Jersey, USA

25. S. Palavajjhalla, R. L. Motard, Joseph, 1994B. Computational Aspects of Wavelets and Wavelet Transforms, In: Wavelet Applications in Chemical Engineering, Motard, R. L. &Joseph, B., (Eds.), 33 83Cluwer Academic Publisher, 0-79239-461-5Massachusetts, USA.

26. A. Petek, V. Doleček, V. Vlachy, 1997Stochastic Analysis of Current Fluctuations During General Corrosion of Stainless Steel in Sulfuric Acid, Corrosion, 53 12 928 934 0010-9312

27. A. Petek, V. Doleček, 2001Interpretation of current noise generation by a simple model, Materials and Corrosion, 52 6 426 429 1521-4176Planinšič, P., Gergič, B., Gleich, D. & Čučej, Ž. (2001). Fuzzy control of subband codded image quality using standard and fuzzy quality measure, Optical Engineering, Vol.8, No.40, (August 2001), pp. 1529-1544, ISSN 0091-3286

28. P. Planinšič, A. Petek, 2003Analysis of electrochemical noise signals with classical methods and methods based on fractal-like wavelets, Proceedings IEEE ICIT 2003 871 876 0-78037-853-9Slovenia, December 10-12, 2003

29. P. Planinšič, A. Petek, 2004Analysis of electrochemical noise signals using fractal-like function wavelets, In: Interdisciplinary applications of fractal and chaos theory, R. Dobrescu, C. Vasilescu, (Ed.), 322 334Editura Academica Romane, 9-37271-070-5Romania

30. P. Planinšič, A. Petek, 2007Electrochemical noise signals discrimination using wavelet-based fractal analysis, Proceeding of 3rd international symposium on Interdisciplinary approaches in fractal analysis IAFA 2007, 322 334 8-42650-823-3Bucaresti, Romania, May 23-25, 2007

31. P. Planinšič, A. Petek, 2008Characterization of corrosion processes by current noise-based fractal and correlation analysis, Electrochimica Acta, 53 16June 2008), 5206 5214 0013-4686

32. M. Sekine, T. Tamura, M. Akay, T. Fujimuto, T. Togawa, Y. Fukuji, 2002Discrimination of Walking Paterns Using Wavelet-Based Fractal Analysis, IEEE Transaction on Neural Sysems and rehabilitation Engineering, 6 3September 2002), 188 196 1534-4320

33. Y. Y. Shi, Z. Zhang, F. H. Cao, J. Q. Zhang, 2008Dimensional analysis applied to pitting corrosion measurements, Electrochimica Acta, 53 6 2688 2698 0013-4686

34. M. Schroeder, 1991 Fractals, Chaos, Power Laws (Minutes from an Infinite Paradise), W. H. Freeman and Company, 0-71672-136-8York, USA

35. J. Smulko, K. Darowicki, A. Zielinski, 2002Pitting Corrosion in Steel and Electrochemical Noise Intensity, Electrochemistry Communication, 4 5May 2002), 388 391 1388-2481

36. A. Wharton, R. J. K. Woo, B. G. Mellor, 2003Wavelet analysis of electrochemical noise measurements during corrosion of austenitic and superduplex stainless steels in chloride media, Corrosion Science, 45 1January 2003), 97 122 0001-0938X

37. G. W. Wornell, 1996 Signal Processing with Fractals, A Wavelet-Based Approach, Prentice Hall PTR, ISBN 013120999X, Upper Saddle River, New Jersey, USA

38. T. Zhang, Y. Shao, G. Meng, F. Wang, 2007Electrochemical noise analysis of the corrosion of AZ91D magnesium alloy in alkaline chloride solution, Electrochimica Acta, 53 17 561 568 0013-4686

Chapter 10

Applications of Discrete Wavelet Transform in Optical Fibre Sensing

Allan C. L. Wong[1] and Gang-Ding Peng[1]

[1] School of Electrical Engineering and Telecommunications University of New South Wales, Australia

1. INTRODUCTION

This chapter presents a comprehensive review of recent advances in the applications of discrete wavelet transform (DWT) in optical fibre sensing. DWT, like Fourier transform (FT), is a versatile and powerful mathematical tool to process, extract and analyse data. In fibre sensing, DWT is particularly useful in the demodulation, demultiplexing and denoising of sensor data, as well as in the detection, extraction and interpretation of measurand-induced change from an acquired sensor signal.

Both continuous and discrete has found a wide variety of applications in fibre sensing, and has been extensively used for fibre Bragg gratings (FBGs) and interferometric sensors. For example, Chan et al. (2007, 2010) used wavelet transform (both continuous and discrete) in reducing noise and increasing wavelength detection accuracy of FBGs; Jones (2000a, 2000b) used in the edge detection and crack detection of FBGs embedded in some structures; Staszewski et al. (1997) and Bang & Kim (2010) used for the detection of acoustic wave induced by impact and defect in composite plates; Gangopadhyay et al. (2005, 2006) used to extract and analyse fibre

Fabry-Perot interferometer signals; Lamela-Rivera et al. (2003) used in the detection of partial discharges from high-power transformers; Tomic et al. (2010) in pressure sensing; and Wang et al. (2001) in the health monitoring of ship hull structure.

This chapter begins with a brief introduction on the applications of DWT in fibre sensing. This follows by the principles and approaches of using DWT. Several important and fundamental formulations and concepts, such as the use of DWT in signal demodulation, demultiplexing and noise reduction will be presented. Next, four exemplary application cases of using DWT in fibre sensing are presented. More specifically, representative topics with regard to sensor signal analysis, signal demodulation, noise reduction and demultiplexing of multiplexed sensor systems are described in details. Finally, conclusions to summarise the chapter is given.

2. PRINCIPLES AND APPROACHES

In fibre sensing, there are two areas of signal analysis that DWT has been widely employed, namely sensor signal demodulation and noise reduction. The former is accomplished by the theory of Multiresolution analysis (MRA) and the latter by the wavelet denoising. This section will describe these two DWT techniques qualitatively, and all of the subsequent examples are essentially based on them.

2.1. Sensor Signal Demodulation

The DWT demodulation technique presented here is used to demodulate and demultiplex many types of multiplexed sensor signals, and it works best with periodic sinusoidal functions, which is often the case for most interferometric signals. Excellent references in the theory and properties of wavelets can be found in (Daubechies, 1992; Mallet, 1998; Sidney Burrus et al., 1998; Vetterli & Kovacevic, 1995). In this section, only the concepts and results that are essential to the understanding of this demodulation technique are included. The principle of operation is based on the theory of MRA, a representation of DWT from a digital signal processing perspective (Mallet, 1989). In the MRA, the sensor signal can be represented as a wavelet series,

$$I(\lambda) = \sum_{m} c_J(m)\phi_{J,m}(\lambda) + \sum_{m}\sum_{j=1}^{J} d_j(m)\psi_{j,m}(\lambda)\,, \tag{1}$$

Where j,m∈Zj,m∈Z, and the integer *J* sets the highest decomposition level.

$$c_j(m) = \langle I(\lambda), \phi_{j,m}(\lambda)\rangle = \int I(\lambda)\varphi^{*}_{j,m}(\lambda)d\lambda \tag{2}$$

are the *j*th-level DWT approximation coefficients, and

$$d_j(m) = \langle I(\lambda), \psi_{j,m}(\lambda)\rangle = \int I(\lambda)\psi^{*}_{j,m}(\lambda)d\lambda \tag{3}$$

are the *j*th-level DWT detail coefficients. * denotes the operation of complex conjugation. $\phi_{j,m}(\lambda) = 2^{j/2}\phi(2^j\lambda - m)$ is the scaling function, and $\psi_{j,m}(\lambda) = 2^{j/2}\psi(2^j\lambda - m)$ is the wavelet function. The DWT coefficients can be computed by a multistage two-channel quadrature mirror filter bank (QMFB). This is formed by the scaling function acting as a low-pass filter, i.e., $\phi(\lambda) = \sum_m h(m)\sqrt{2}\phi(2\lambda - m)$, where $h(m),\ m \in Z$ are the low-pass filter coefficients. The complementary wavelet function acts as a high-pass filter, i.e., $\psi(\lambda) = \sum_m g(m)\sqrt{2}\phi(2\lambda - m)$, where *g*(*m*) are the high-pass filter coefficients. The filter coefficients are related by $g(m) = (-1)^m h(1-m)$ In other words, in the QMFB the scaling and wavelet functions simultaneously perform the low-pass and high-pass filtering on the sensor signal. At the *j*th-stage of the QMFB, the DWT approximation coefficients are given by,

$$A_j = c_j(m) = \sum_k h(k - 2m)c_{j-1}(k)\,, \tag{4}$$

and the DWT detail coefficients are given by,

$$D_j = d_j(m) = \sum_k g(k - 2m)c_{j-1}(k)\,. \tag{5}$$

Then, Eq. (1) can be expressed in the form,

$$I(\lambda) = \sum_{m} A_J + \sum_{m}\sum_{j=1}^{J} D_j\,. \tag{6}$$

From this representation, the *j*th-level DWT coefficients can be computed by convolving the QMFB with the *(j–1)*th-level approximation coefficients, i.e., Aj−1=Aj+DjAj−1=Aj+Dj. The schematic diagram of the multistage decomposition operation using the QMFB is shown in Fig. 1. By repeating this decomposition in cascade using the most recent *A j* as inputs, we can compute the DWT coefficients to the resolution level of interest. The usefulness of the MRA representation is that a signal can be decomposed into different levels of wavelet coefficients according to the frequency components that comprised the whole signal. The higher the wavelet levels, the lower the frequency components of the signal remain. Fig. 2 shows graphically the frequency representation of a signal being decomposed into various wavelet levels. *f c* is the centre frequency of the whole spectral bandwidth of the signal. If the multiplexed sensor signal is tailored in such a way that each sensor has a different (signal) frequency, after this cascaded operation, their original sensor signals will appear on different wavelet levels.

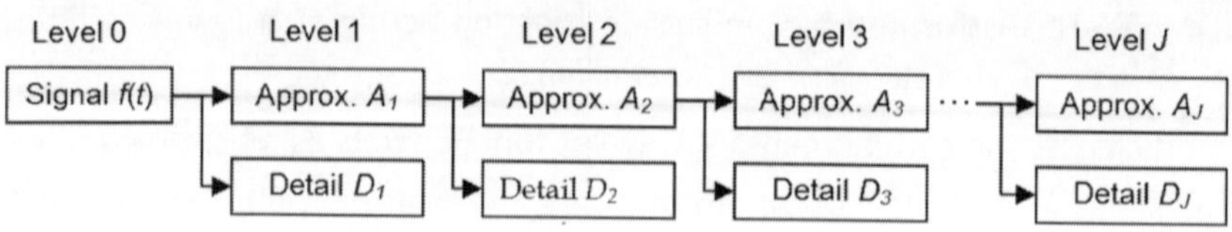

Figure 1. Multistage decomposition of a signal using QMFB.

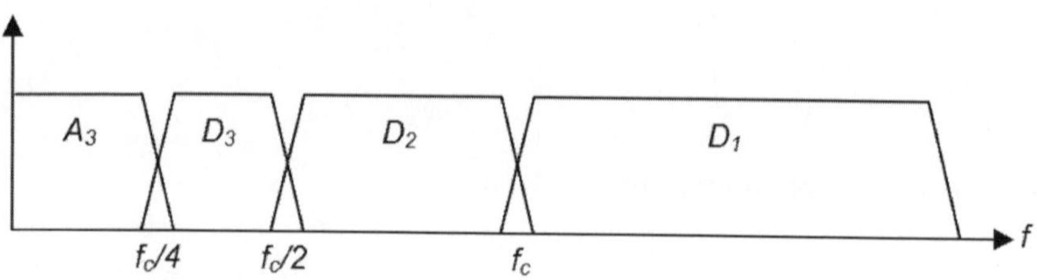

Figure 2. Frequency representation of a signal decomposed into various wavelet levels.

2.2. Wavelet Denoising

Wavelet denoising (Donoho, 1994) is a nonlinear noise reduction technique based on the DWT to remove/reduce the noise in the signal while preserving the overall signal features. The generic denoising algorithm is depicted in Fig. 3. In principle, wavelet denoising attempts to decompose a signal using the DWT to obtain the wavelet coefficients, and then apply a thresholding method or a shrinkage rule to each wavelet coefficient. The threshold can be obtained by using some risk estimators or by empirically finding a value. The method then either keeps or shrinks all wavelet coefficients that are above the threshold, and suppresses all those below the threshold. Then, the signal is reconstructed by taking the inverse DWT with the noisy part being removed. Here, two denoising techniques that are frequently employed in fibre sensing are discussed, namely hard thresholding (HT) and block-level thresholding (BLT). The former is a simple and effective method, while the latter is automatically incorporated with the DWT demodulation technique, making it a very attractive and convenient denoising method.

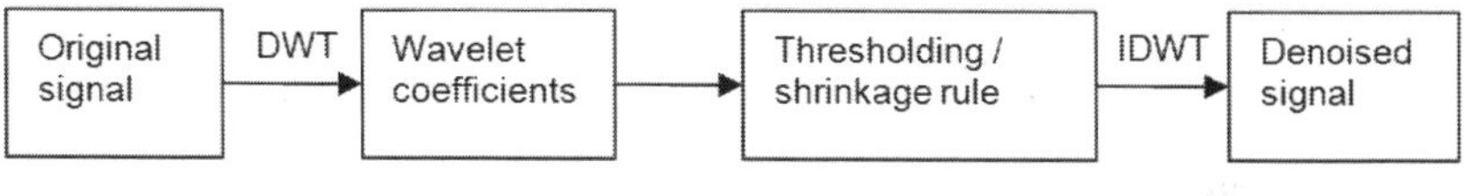

Figure 3. Generic wavelet denoising algorithm.

2.2.1. Hard Thresholding

When a sensor signal is acquired from a measurement system, the signal consists of broadband noise caused by the quality of the sensor, as well as the quantisation and finite resolution errors of the equipment. It is known that broadband noise is difficult to remove using conventional filtering without altering the signal. But this can be done by the HT wavelet denoising. Let the sensor signal intensity be represented as a wavelet series,

$$I(\lambda) = \sum_{m,n=-\infty}^{\infty} \alpha_{m,n}\psi_{m,n}(\lambda), \tag{7}$$

where ψm,n $\psi_{m,n}(\lambda) = 2^{m/2}\psi(2^m\lambda - n)$, $\psi \in L^2(\Re)$, $m, n \in Z$ are the orthonormal wavelet basis functions. By taking the DWT, the wavelet coefficients are,

$$\alpha_{m,n} = \langle \psi_{m,n}(\lambda), I(\lambda) \rangle = \int_{-\infty}^{\infty} \psi^*_{m,n}(\lambda) I(\lambda) d\lambda \,. \tag{8}$$

The HT then sets a threshold value, *h*, to the wavelet coefficients, such that any coefficients below the threshold are suppressed, whereas the coefficients above the threshold are retained. Since the noise components are usually of low magnitude, the threshold can be set to a value high enough to eliminate the noisy coefficients, and low enough to retain useful signal coefficients. The HT operation can be represented by (Wong et al., 2005),

$$\xi_h(\alpha_{m,n}) = \hat{\alpha}_{m,n} = \begin{cases} \alpha_{m,n}, & |\alpha_{m,n}| > h \\ 0, & |\alpha_{m,n}| \le h \end{cases} \,. \tag{9}$$

Since the HT is applied to the entire spectral range of the sensor signal, the broadband noise can be effectively removed. Once the noisy coefficients are removed, the signal can be reconstructed by taking the inverse DWT,

$$I'(\lambda) = \sum_{m,n=-\infty}^{\infty} \hat{\alpha}_{m,n} \psi_{m,n}(\lambda) \,. \tag{10}$$

From another perspective, suppose a sensor signal consists of an ideal noise-free part *f(λ)* and a noisy part *σ(λ)* with noise level *p*, i.e., $I(\lambda) = f(\lambda) + p\sigma(\lambda)$. If we set *h* = *p*, i.e., setting the threshold equal to the noise level, then the HT will suppress the noisy coefficients to zero while retaining the noise-free coefficients unaffected. The ideal case is to have| $|I(\lambda) - f(\lambda)| \to 0$, i.e., ., the reconstructed signal is as close as practicable to the ideal, noise-free signal.

2.2.2. Block-Level Thresholding

The BLT can be considered as a modified version of the block-thresholding method (Wong et al., 2006b). In the block-thresholding method, instead of applying a threshold to each wavelet coefficient term-by-term (as is for HT), the threshold is applied to a block of wavelet coefficients, with the threshold value determined by calculating and minimising the risk using the block James-Stein estimator. On the other hand, the BLT method attempts to set an entire level of wavelet coefficients as a block, such that the entire level is either retained or discarded. There is no need to estimate the risk or to find a value empirically in order to obtain the threshold value. For example, if the sensor signal has Gaussian white noise associated with it, such that the noise components spanned the whole frequency range of the signal. From the theory of MRA, the 1st-level detail coefficients account for the upper-half of the whole frequency range of the original signal, the 2nd-level account for the upper-half frequency range of the 1st-level approximation coefficients, and so on (c.f. Fig. 2). Therefore, the first two levels of detail coefficients cover 75 % of the whole frequency range of the sensor signal, and the noise components within this range will be removed if we discard the detail coefficients at these levels. For the signal expressed in the form of Eq. (6) with levels j=1,2,...,Jj=1,2,...,J, the BLT operation for a chosen level of approximation coefficients ($j = J A$) used to demodulate the sensor signal is given by,

$$BLT(j = J_A) \Rightarrow I(\lambda) = \begin{cases} \sum_m A_{J_A}, & \text{for } j = J_A \\ 0, & \text{otherwise} \end{cases}, \tag{11}$$

Similarly, for a chosen level of detail coefficients ($j = J D$) used to demodulate the sensor signal, the BLT operation is given by,

$$BLT(j = J_D) \Rightarrow I(\lambda) = \begin{cases} \sum_m D_{J_D}, & \text{for } j = J_D \\ 0, & \text{otherwise} \end{cases}. \tag{12}$$

In other words, the BLT discards all the levels of detail coefficients that are at (for approximation coefficients) or below (for detail coefficients) the desired level chosen to be used for sensor signal demodulation. Additionally, the BLT denoising technique can be applied as a standalone technique for any type of signal. In a more general form, the BLT for a chosen level j_{BLT}, where $1 \leq j_{BLT} \leq J$, is given by (Wong et al., 2007a),

$$BLT(j_{BLT}) \Rightarrow I(\lambda) = \begin{cases} \sum_m A_J + \sum_m \sum_{j=j_{BLT}+1}^{J} D_j, & \text{for } j_{BLT} < j \leq J \\ \sum_m A_J, & \text{for } j = j_{BLT} \\ 0, & \text{for } 1 \leq j < j_{BLT} \end{cases}. \tag{13}$$

An inverse DWT is then applied to the wavelet coefficients to reconstruct the signal with the noisy part removed. The choice of *j BLT* requires the signal to preserve its characteristic features for demodulation purposes, while being able to maximally remove the broadband noise components associated with it.

In the subsequent sections, four exemplary applications, which are the research works conducted by the authors, will be presented to demonstrate the power and versatility of the DWT.

3. APPLICATION CASE 1 – SIMULTANEOUS MULTI-SENSOR SIGNAL DEMODULATION

In the first example, a new simultaneous demodulation technique was proposed for a multiplexed fibre Fizeau interferometer (FFI) and FBG sensor system, based on the DWT technique described in the last section (Wong et al., 2006a, 2006b). In relation to this demodulation technique, the BLT denoising technique was applied automatically and simultaneously to reduce the noise associated with the sensor signal. It is known that FFIs and FBGs are two of the most widely studied types of fibre-optic sensors (Lee, 2003). They have distinct measurand-induced responses and dynamic ranges, and hence, when they are combined together, they can measure different measurands simultaneously. This demodulation technique outperforms currently reported techniques, and the key advantages are: (i) it overcomes the disadvantages associated with interferometric methods; (ii) it only requires a simple setup to interrogate and multiplex the sensors. All the data acquisition, signal processing, and calculations are carried out digitally by a computer program, and no complicated demodulating electronics are needed; (iii) it is a completely passive system that does not require any mechanical moving parts or active modulation, making it especially suited for continuous long-term quasi-static sensing; (iv) and it automatically reduces the signal noise through the BLT, without the need of any additional filtering techniques.

The FFI and FBG sensors are multiplexed using a hybrid of spatial-frequency-division multiplexing (SFDM) and wavelength-division multiplexing (WDM). Specifically, SFDM is used to multiplex the FFI sensors, in that each sensor produces a sinusoidal interference pattern with the spatial frequency specified by the cavity length. The multiplexed FFIs signal can then be demodulated using the FT peak detection method, which will be described later. For FBG sensors, WDM is employed in order to take the advantage of the wavelength-encoded nature. In the wavelength domain, each FBG has a narrowband Bragg peak in the reflection spectrum. As a result, the multiplexed signal is of two extremes: the FBG signal (i.e., the main peak) is localised with compact support in the original (wavelength) domain, but spans across the entire spectral bandwidth in the dual (spatial-frequency domain). The opposite case applies to the FFI signal. That is, in the original domain the signal is spanned along the spectral bandwidth of the light source. But in the dual domain, the signal is localised with compact support, and is shown as a sharp peak. In this case, the multiplexed signal can be easily separated by the DWT technique and individual sensor signals will appear on two distinct wavelet levels.

Fig. 4(a) shows the experimental setup for the sensor system, An amplified spontaneous emission (ASE) light source was connected to a 3-dB fibre coupler to illuminate the FFI and FBG sensors that are serially multiplexed. This reflected signal was acquired by an optical spectrum analyser (OSA) and transferred to a computer. The multiplexed signal is the superposition of the two individual sensor signals. Assuming no insertion loss, for *N* FFIs and *M* FBGs, the measured spectrum is of the form (Wong et al., 2006b):

$$\begin{aligned} I(\lambda) &= A(\lambda)\left(\sum I(\lambda)_{FFI} + \sum I(\lambda)_{FBG}\right) \\ &= A(\lambda)\left\{\sum_{k=1}^{N} r_{1,k} r_{2,k}\left[1 + V_k S(\lambda)\cos(4\pi d_k / \lambda)\right] + \sum_{l=1}^{M} R_{B,l}\exp\left[-(\lambda - \lambda_{B,l})^2 / 2\delta\lambda_{B,l}^2\right]\right\} \end{aligned} \quad (14)$$

where *A(λ)* is the incident light source intensity, *r i* are the reflection coefficients of the two fibre ends that form the FFI cavity, *d* is the cavity length, *V* is the fringe visibility, *S(λ)* is the erbium-doped fibre (EDF) amplifier spectral profile, *λ B* is the centre Bragg wavelength, *R B* is the peak reflectivity of the FBG, and *δλ B* is a parameter related to the FBG bandwidth. A typical multiplexed signal for *N* = *M* = 1 is shown in Fig. 4(b). From the figure, the FBG can be identified by the distinct narrowband peak (around 1533 nm), whereas the FFI produced a sinusoidal interference pattern that spanned the light source bandwidth, and superimposed on its spectral profile.

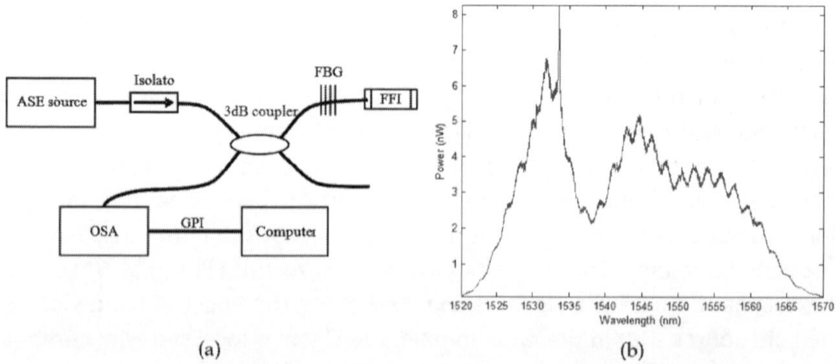

Figure 4. a) Experimental setup for the multiplexed FFI and FBG sensor system; and (b) a typical multiplexed sensor signal.

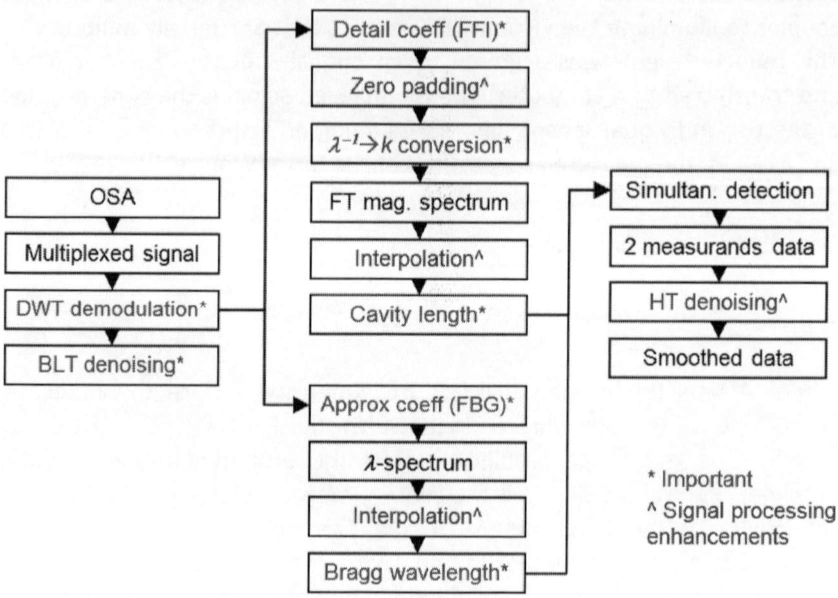

Figure 5. Simultaneous demodulation algorithm with wavelet denoising.\

The simultaneous demodulation algorithm is depicted in Fig. 5. For a multiplexed signal retrieved from the OSA, the DWT was applied to decompose it into multilevel wavelet coefficients. In parallel to the demodulation, the BLT was applied simultaneously such that the decomposed signal was denoised. The discrete Meyer wavelet was chosen to be the kernel function for DWT because the number of vanishing moments and regularity are suitable for analysing this type of multiplexed signal. For the FBG sensor, since the received signal from the OSA was a function of wavelength, the Bragg wavelength can be determined directly from a level of approximation coefficients, which effectively gave the denoised version of the original spectrum. The BLT technique automatically discarded the wavelet levels that were below the chosen level of approximation coefficients for the FBG signal, and thus reduced the signal noise without explicitly applying any additional filtering technique. Then, an interpolation utilising the piecewise-continuous cubic-spline function was applied to the Bragg peak to increase the wavelength resolution. Next, the cavity length of the FFI was determined. After taking the DWT, the detail coefficients were indeed the original interference pattern of the FFI, which was then demodulated using the FT peak detection method. First, the signal was Fourier transformed and each sensor had its own peak in the magnitude spectrum. The important step was to convert the variables of the detail coefficients from wavelength to wavenumber. In doing so, the FT dual-domain variable was related to the sensor cavity length. The cavity length can then be obtained directly from the location of the amplitude peak in the magnitude spectrum (Wong et al., 2005). Once the characteristic change of both sensors are known, i.e., Bragg wavelength of the FBG and cavity length of the FFI, simultaneous measurement of two measurands can be achieved by monitoring the changes of both sensors. In order to do so, the measurands-induced responses, i.e., the elements in the sensitivity matrix need to be determined. The HT denoising technique (described in previous section) can be used to denoise and smooth out the measured data. Thus, this demodulation technique provided a complete process for a practical sensor system from the acquisition of raw data to human understandable measurand outputs, and all steps were performed digitally by a computer program.

Having described in details the demodulation algorithm, for an acquired multiplexed signal [Fig. 4(b)], after taking the DWT, the approximation coefficients that represent the FBG signal is shown in Fig. 6(a). The approximation coefficients give a smoothed and noise-reduced version of the original noisy signal. Fig. 6(b) shows the magnification around the Bragg peak, and the Bragg wavelength shift can be found directly by employing a wavelength detection method. A resolution of ~1.2 pm after interpolation was achieved. For the FFI sensor, after taking the DWT, Fig. 7(a) shows the detail coefficients, which effectively represented the mean-removed and windowed version of the original interference pattern. Fig. 7(b) shows the

Fourier transformed magnitude spectrum of the detail coefficients around the amplitude peak. It can be seen that, after conversion of x-variable, the cavity length can be directly determined from its position in the magnitude spectrum. A resolution of ~0.04 µm was achieved after interpolation.

An application of this demodulation technique was demonstrated through the simultaneous measurement of strain and temperature of an aluminium (Al) plate. With the setup from Fig. 4(a) the FBG was loosely adhered onto the Al-plate, such that it was not affected by the strain field. The FFI adhered firmly onto the Al-plate next to the FBG using some superglue, and so it was affected by both strain and temperature changes of the plate. Before proceeding to the actual experiment, the coefficients in the sensitivity matrix needed to be determined. To illustrate this, let the two sensors be represented as,

$$\begin{aligned}\Delta\zeta_1 &= K_{1\varepsilon}\Delta\varepsilon + K_{1T}\Delta T \\ \Delta\zeta_2 &= K_{2\varepsilon}\Delta\varepsilon + K_{2T}\Delta T\end{aligned}, \tag{15}$$

where $\Delta\zeta\ i$ are the measurand-induced physical change of the sensors, and $K\ ij$ are the measurand-induced responses (i.e., sensitivity coefficients). $\Delta\varepsilon$ and ΔT are the strain and temperature changes, respectively. Eq. (15) is a set of simultaneous linear equations, which can be expressed in matrix form as,

$$\begin{pmatrix}\Delta\zeta_1 \\ \Delta\zeta_2\end{pmatrix} = \begin{pmatrix}K_{1\varepsilon} & K_{1T} \\ K_{2\varepsilon} & K_{2T}\end{pmatrix}\begin{pmatrix}\Delta\varepsilon \\ \Delta T\end{pmatrix}. \tag{16}$$

This equation can be solved for the measurand vector by inverting the *K*-matrix, i.e.,

$$\begin{pmatrix}\Delta\varepsilon \\ \Delta T\end{pmatrix} = \frac{1}{\det(K)}\begin{pmatrix}K_{2T} & -K_{1T} \\ -K_{2\varepsilon} & K_{1\varepsilon}\end{pmatrix}\begin{pmatrix}\Delta\zeta_1 \\ \Delta\zeta_2\end{pmatrix}, \tag{17}$$

Where $\det(K) = K_{1\varepsilon}K_{2T} - K_{1T}K_{2\varepsilon} \neq 0$. . This shows the strain and temperature can be simultaneously separated and measured provided the elements in the square matrix are pre-determined. This can be done by measuring one measurand (while the other kept constant) at a time against the spectral response of each sensor. For this sensor arrangement, with the details described in (Wong et al., 2006b), the strain and temperature responses of the FBG and FFI sensors are shown in Figs. 8 and9, respectively. From the linear regression fits, for the FBG sensor the strain and temperature sensitivity coefficients are 1.2 pm/με and 10.4 pm/°C, respectively; whereas for the FFI sensor, the respective sensitivity coefficients are 4.06×10^4 pm/με and 1.12×10^6 pm/°C. Since the FBG was not affected by strain change, the system of linear equations is given by,

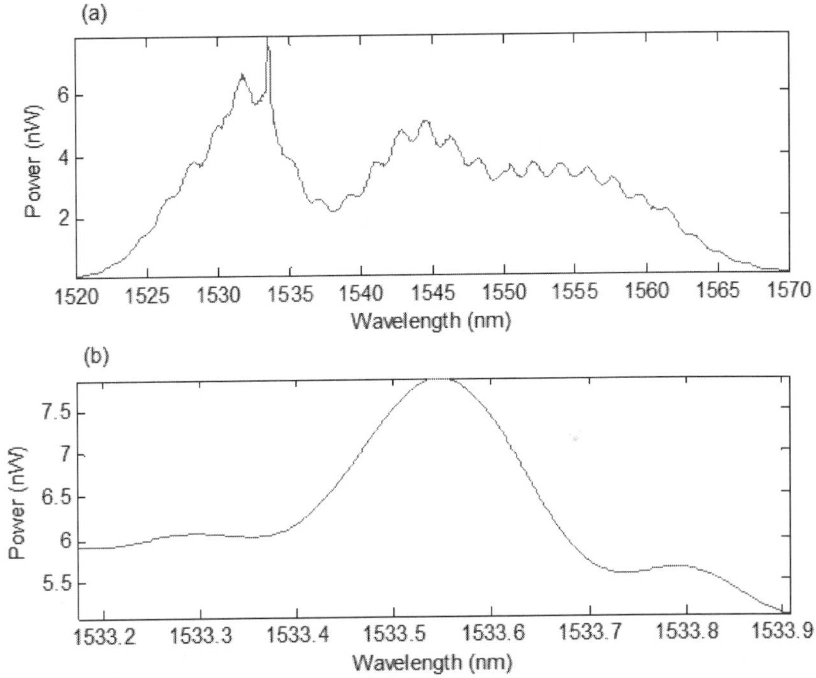

Figure 6. a) Approximation coefficients for the FBG; (b) around the Bragg peak.

$$\begin{pmatrix} \Delta d \\ \Delta\lambda_B \end{pmatrix} = \begin{pmatrix} 4.06\times10^4 & 1.12\times10^6 \\ 0 & 10.4 \end{pmatrix} \begin{pmatrix} \Delta\varepsilon \\ \Delta T \end{pmatrix}, \qquad (18)$$

and

$$\begin{pmatrix} \Delta\varepsilon \\ \Delta T \end{pmatrix} = \begin{pmatrix} 2.5\times10^{-5} & -2.7 \\ 0 & 9.6\times10^{-2} \end{pmatrix} \begin{pmatrix} \Delta d \\ \Delta\lambda_B \end{pmatrix}, \tag{19}$$

where Δd and $\Delta\lambda$ *B* are the FFI cavity length change and FBG Bragg wavelength shift, respectively. The sensors were left in the laboratory overnight for a period of 14 hours, with no external axial strain applied. A LabVIEW program was written to acquire the sensor signal in real-time at a rate of ~6 s per reading. The FFI subjected to both the thermal strain and temperature change of the Al-plate, while the FBG subjected to temperature change only. By using Eq. (18), the separated effects of thermal strain and temperature change over the measured period are shown in Fig. 10. Although both sensors experienced the same temperature change, the thermal strain of the Al-plate did not correlate very well with its temperature change. This is because the FBG was loosely adhered onto the Al-plate, it was influenced by the surrounding environment and temperature change more than the thermal response of the Al-plate.

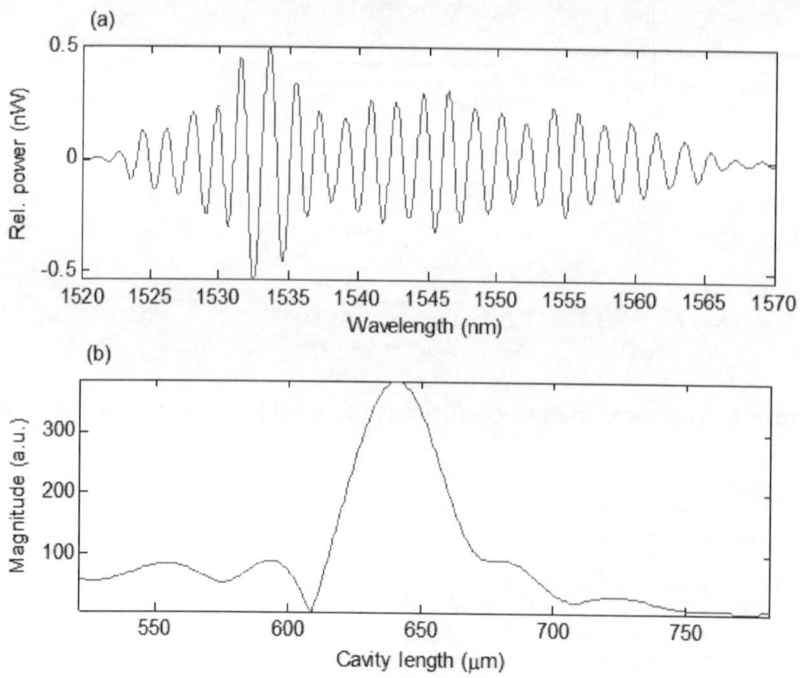

Figure 7. a) Detail coefficients for the FFI; (b) FT magnitude spectrum around the peak.

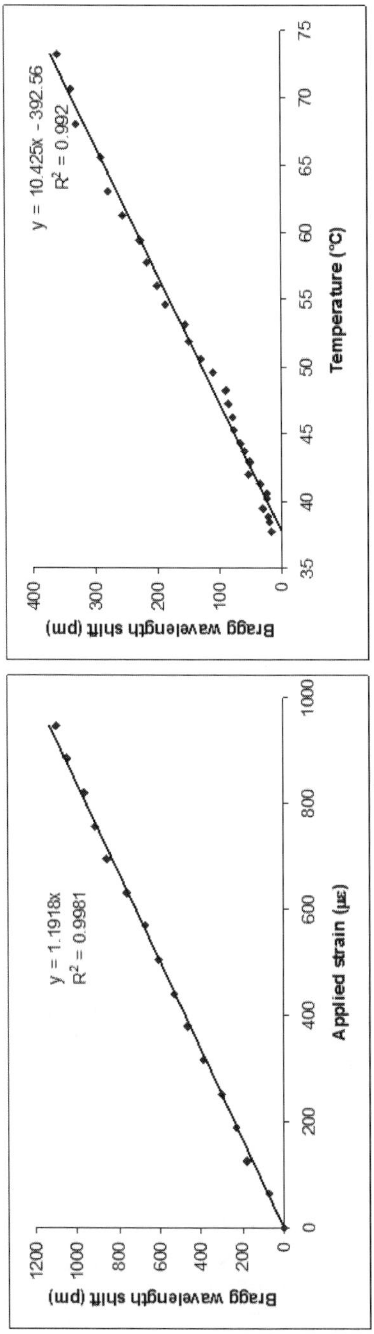

Figure 8. a) Strain and (b) temperature responses of the FBG sensor.

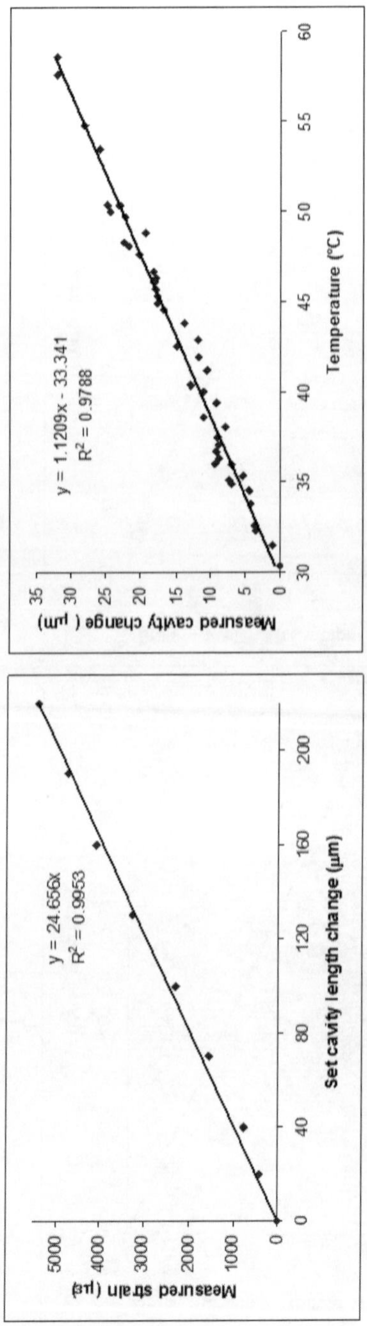

Figure 9. a) Strain and (b) temperature responses of the FFI sensor.

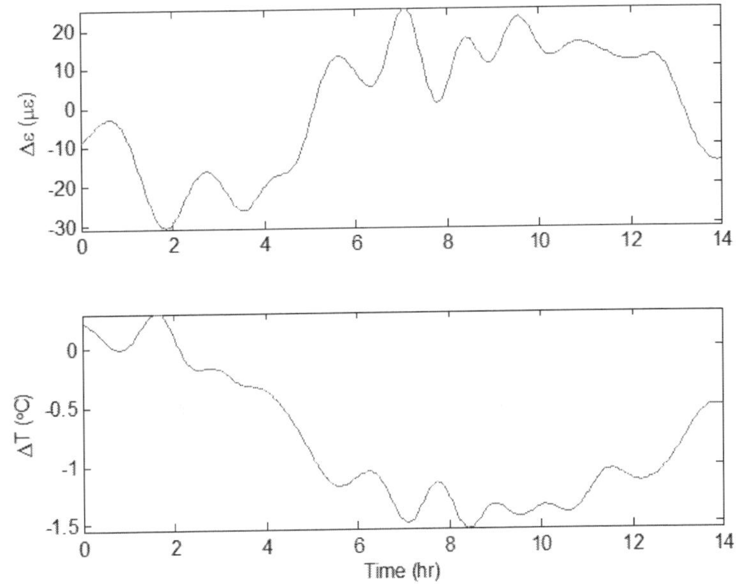

Figure 10. Simultaneous measurement of strain and temperature of Al-plate.

4. APPLICATION CASE 2 – MULTIPLEXING AND DEMULTIPLEXING OF MULTI-SENSORS

In the second example, a new type of FBG called amplitude-modulated chirped FBGs (AMCFBGs) was designed and fabricated, and based on that, a new multiplexing technique called spectral overlap multiplexing was proposed and experimentally demonstrated (Childs et al., 2010, Wong et al., 2007a,2007b, 2010). We show that, the DWT played a central role in the demultiplexing, demodulation and analysis of these multiplexed novel sensor signals. The AMCFBGs are similar to ordinary chirped FBGs, i.e., they have a broad and flat reflection spectrum. The subtle difference is the addition of an amplitude-modulation function to the refractive index modulation of the fibre core, which is achieved by varying the induced DC refractive index during the writing process. With this index modulation, the reflection spectrum has a unique signature – a sinusoidal modulation on its flat-top region. The amplitude-modulation function for the refractive index modulation is given by $f_d(z) = w \cdot \sin^2(\pi dz / L_g)$, where *w* is an apodisation function such as a raised-cosine function, *d* is the number of periods (frequency) of modulation, and *L g* is the length of the grating. The reflection spectrum for one AMCFBG can be approximated by the expression (Wong et al., 2007a),

$$R(\lambda) \approx \chi_{(\lambda_{\min}, \lambda_{\min} + c_L L_g)}(\lambda)\left[R_{\min} + (R_{\max} - R_{\min}) f_d\left(\frac{\lambda - \lambda_{\min}}{c_L}\right)\right], \tag{20}$$

where *λ min* is the initial wavelength of the grating, *R min* and *R max* are the reflectivities of the troughs and peaks of the grating spectrum, respectively, *c L* is the linear chirp rate of the phase mask used for writing the gratings, and *I* (*x*) is the characteristic function on the interval *I* which equals 1 when *x* is an element of *I* and zero otherwise. The fabrication procedure can be found in (Wong et al., 2007b). As an example of an AMCFBG that we fabricated, Fig. 11(a) shows the amplitude-modulation function with a raised-cosine apodisation, *d* = 5 and *L g* = 10 mm; and Fig. 11(b) shows the corresponding measured reflection spectrum.

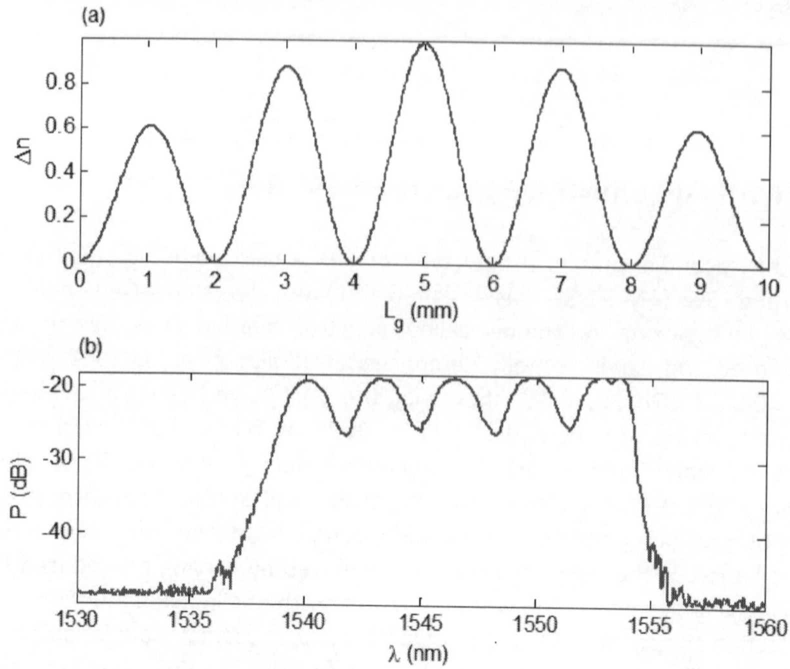

Figure 11. a) Index modulation function used in the fabrication; (b) measured reflection spectrum of the AMCFBG.

In the wavelength domain, ordinary FBGs cannot be easily distinguished if they are spectrally overlapped. Using the amplitude-modulation as a unique signature for each grating, the AMCFBGs are able to completely overlap one another having the same spectral characteristics, i.e., centre Bragg wavelength, bandwidth and reflectivity, yet still be uniquely distinguishable from each other. The uniqueness of each AMCFBG is defined by the number of periods (spatial-frequency) of its amplitude-modulation, and for a set of spectrally-overlapped gratings, no two gratings can have the same number of periods. With such unique signatures for the AMCFBGs, viable methods are needed to demultiplex and demodulate the multiplexed signals, and that was accomplished by using the DWT technique similar to that applied in the previous section. This is the basis of the new spectral overlap multiplexing technique. Since this multiplexing is fully compatible with the WDM, the sensor count can potentially be increased by several folds.

To demonstrate the spectral overlap multiplexing using AMCFBGs, we performed two experiments. In the first experiment, two spectrally overlapped gratings, S1 and S2, were employed, and the schematic diagram of the setup is depicted in Fig. 12. The individual spectra of S1 and S2, as well as the combined spectra are shown in Fig. 13 (a) – 13 (c), respectively. It is clear that, WDM scheme would not allow such overlapping, and without a suitable signal processing technique, it is unlikely the multiplexed signal be separated and analysed. However, by taking the DWT, the unique signatures, i.e., the sinusoidal modulations of the gratings, can be recovered. Fig. 14 shows that the detail coefficients of the two gratings, and the unique modulated periods are unambiguously identified. That is, the multiplexed signal has been successfully demodulated whereby the measurand-induced wavelength shifts can be measured individually by tracking the 'phase-shift' of the modulations. The strain response and crosstalk of the multiplexed AMCFBGs under this multiplexing scheme were conducted. A total of ten 20-cent coins (each weighs 11.3 g, which corresponds to a strain of 125 με) were applied to S1, while S2 was left unstrained. For each coin applied, ten readings were taken. Fig. 15 (a) shows the measured strain responses of S1 and S2 as a function of applied strain. The strain values were obtained by using the DWT technique mentioned before. The lines are the ideal strain values. It can be seen that S1 measured the strain very accurately. For strains applied up to 1250 με, the absolute error is within ±20 με as shown in Fig. 15 (b). The corresponding strain response of S2 showed little crosstalk, with a maximum value of about 16 με as shown in Fig. 15 (b). This small amount of crosstalk implies that adverse effects, such as spectral shadowing, did not impose much of a problem on this multiplexing technique.

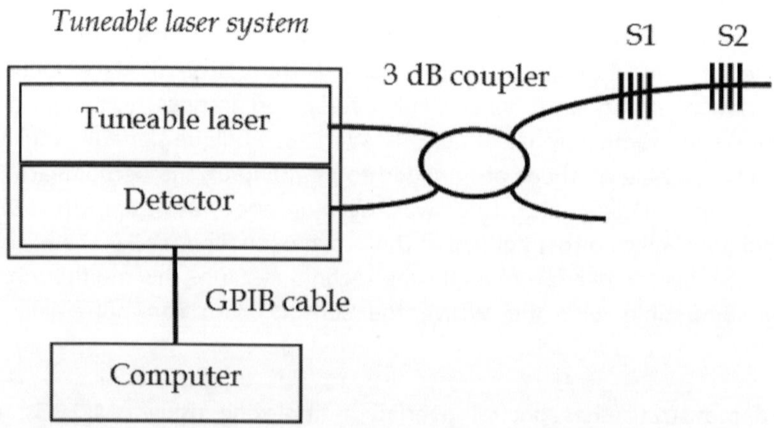

Figure 12. Experimental setup of the multiplexed AMCFBG system.

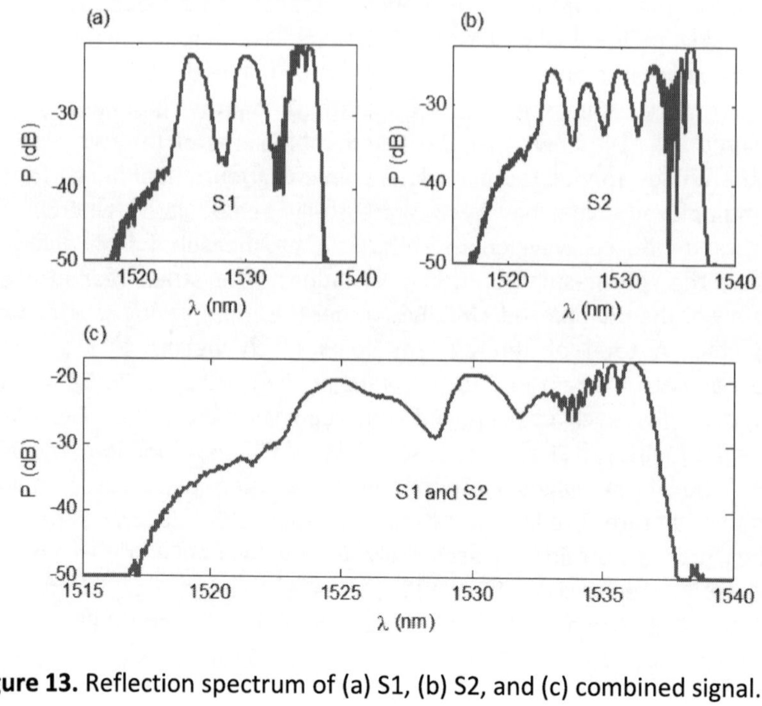

Figure 13. Reflection spectrum of (a) S1, (b) S2, and (c) combined signal.

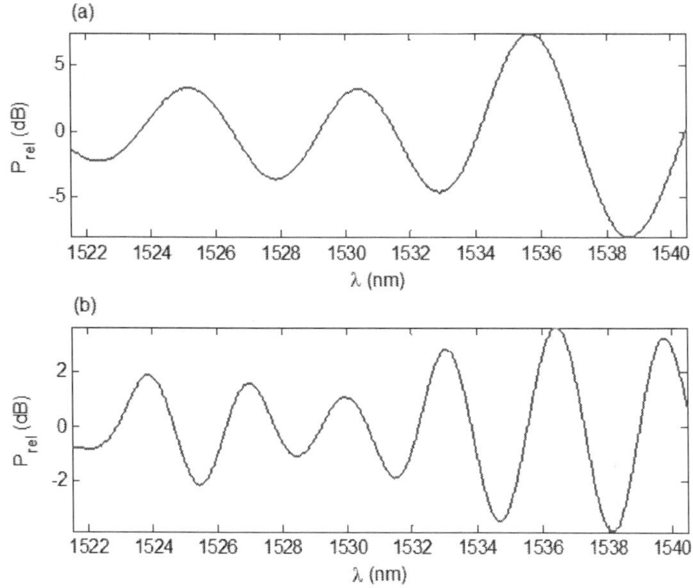

Figure 14. Detail coefficients for (a) S1 and (b) S2 gratings.

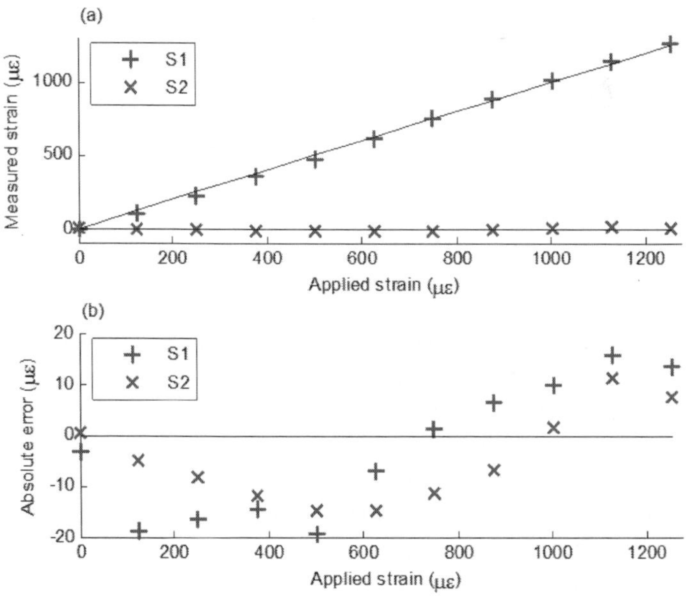

Figure 15. a) Strain measurements and (b) absolute errors of S1 (strained) and S2 (unstrained).

In the second experiment, simultaneous strain and temperature measurement of an Al-alloy plate was carried out. The experiment was similar to that in the previous section. That is, the AMCFBG sensors were first characterised to obtain the strain and temperature sensitivity coefficients, then simultaneous two-parameter measurement of the material was performed. Strain and temperature changes were obtained by measuring the wavelength (i.e., 'phase') shifts of the sensors and calculating their corresponding values. Two overlapped AMCFBGs were used, and to effectively utilise the advantage of having the same Bragg wavelength, the reference grating method (Wong et al., 2007b) was used. In such method, the strain sensor was firmly attached onto the structure, while the temperature sensor was placed under the same environmental conditions but unstrained. As such, the former experienced wavelength shifts due to both strain and temperature, while the latter only experienced the shift due to temperature change. With reference to Eq. (16), the set of simultaneous equations is given by,

$$\begin{pmatrix} \Delta\lambda_{B1}/\lambda_{B1} \\ \Delta\lambda_{B2}/\lambda_{B2} \end{pmatrix} = \begin{pmatrix} k_{1\varepsilon} & k_{1T} \\ 0 & k_{2T} \end{pmatrix} \begin{pmatrix} \Delta\varepsilon \\ \Delta T \end{pmatrix}, \text{ or } \begin{pmatrix} \Delta\varepsilon \\ \Delta T \end{pmatrix} = \frac{1}{k_{1\varepsilon}k_{2T}} \begin{pmatrix} k_{2T} & -k_{1T} \\ 0 & k_{1\varepsilon} \end{pmatrix} \begin{pmatrix} \Delta\lambda_{B1}/\lambda_{B1} \\ \Delta\lambda_{B2}/\lambda_{B2} \end{pmatrix}, \tag{21}$$

where the subscripts 1 and 2 represent strain and temperature sensors, respectively. Some approximations are made: (a) as both sensors experience the same temperature change, $k_T = k_{1T} = k_{2T}$; (b) the Bragg wavelengths are the same for both sensors and so $\lambda_B = \lambda_{B1} = \lambda_{B2}$. By setting $k_\varepsilon = k_{1\varepsilon}$ and modifying Eq. (21), the expressions for strain and temperature change are then given by,

$$\Delta\varepsilon = \frac{1}{k_\varepsilon}\frac{(\Delta\lambda_{B1} - \Delta\lambda_{B2})}{\lambda_B}; \text{ and } \Delta T = \frac{1}{k_T}\frac{\Delta\lambda_{B2}}{\lambda_B}. \tag{22}$$

Eq. (22) shows that the strain of a structure under test can be simply obtained from the differential wavelength shift between the two sensors, and temperature from the temperature sensor alone. Both the strain and temperature coefficients can be determined experimentally by measuring their measurand-induced responses. Fig. 16 shows the strain and temperature responses of an AMCFBG. Based on the slopes of the linear regression fits, the strain and temperature sensitivity coefficients were 1.20 pm/με and 10.06 pm/°C, respectively. The temperature measurement was carried out when the sensor was free and unstrained. However, when it was used to measure the Al-alloy plate, due to the thermal expansion mismatch between the plate and the silica fibre, temperature response needed to be re-measured with the sensor adhered onto the plate. The measured temperature coefficient is found to be 38.12 pm/°C. Now, the strain and temperature can be expressed numerically as,

$$\Delta\varepsilon = 0.8295 \cdot (\Delta\lambda_{B1} - \Delta\lambda_{B2}) \ ; \ \text{and} \ \Delta T = 0.0262 \cdot \Delta\lambda_{B2}, \tag{23}$$

where the Bragg wavelength shift, strain and temperature change are in units of pm, με and °C, respectively. To practically measure the strain and temperature change of the Al-alloy plate, one of the AMCFBG sensors was adhered firmly onto the plate with epoxy resin, whereas the other sensor was loosely attached such that it was free and unstrained from any strain field of the plate. The plate was placed inside a polyurethane foam box to minimise environmental perturbations. With the setup of Fig. 12, experiment was performed for a period of 18 h. By using Eq. (23), the strain and temperature changes are shown in Fig. 17. From the figure, both the strain and temperature curves changed in a much correlated manner, indicating that the strain was mainly thermally-induced. This is obvious as there was no external strain applied throughout the experiment, and unlike the previous results in Fig. 10, the experiment was isolated and so the thermal strain followed well with temperature change. The net change in strain and temperature during this 18 h period were about 40 με and 1.5 °C, respectively. Since the plate was placed inside a foam box that had a relatively high heat capacity, the change in temperature was a bit small. By plotting a graph of temperature vs. strain, the slope of the linear regression fit gives a thermal strain sensitivity of 33 με/°C.

5. APPLICATION CASE 3 – MULTIPLEXING OF PHOTONIC CRYSTAL FIBRE SENSORS

In the third example, DWT was applied to the multiplexing and demultiplexing of a relatively new class of fibre – photonic crystal fibres (PCFs) (Fu et al., 2009). PCFs distinguish themselves from conventional fibers that they consist of microstructured air holes along the fibre, and a wide variety of cross-section air hole arrangements can be designed to suit different applications. The particular type of PCF used in the example was called polarisation-maintaining PCFs (PM-PCFs) that have the characteristics of high birefringence and low temperature sensitivity, and so is suitable for single parameter sensing where cross-sensitivity issue can be minimised.

However, at present, all reported PCF sensors were operated as single sensors, and a main reason was due to the difficulty in demultiplexing and demodulating the multiplexed PCF sensor signals, even though the multiplexing schemes are simple and easy to implement. To overcome this, we demonstrate the use of DWT to separate the multiplexed sensor signal, such that the change from each individual sensor can be extracted and measured.

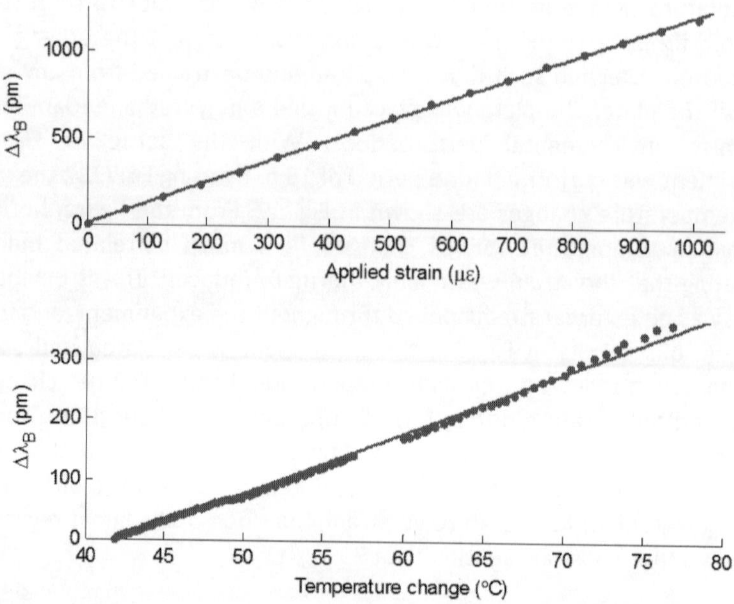

Figure 16. Strain and temperature responses of the AMCFBG.

In our experimental setup, two PM-PCF sensors, PM-PCF1 and PM-PCF2, were multiplexed in series, as shown in Fig. 18. Each sensor unit was arranged in Sagnac interferoemter configuration with a section of PM-PCF as the birefrigent element, and the output signal can be represented by the transmission matrix as, $T=[1-\cos(\delta)]/2T=[1-\cos(\delta)]/2$. The phase difference δ introduced by the PM-PCF with a length of L to the two light beams is wavelength dependent and is given by $\delta=2\pi BL/\lambda\delta=2\pi BL/\lambda$. The period of the output spectrum, i.e., the spacing between two adjacent minima, is $S=\lambda^2/(BL)$, where B is the birefringence of the PM-PCF. The birefringence change due to environmental parameters can then be detected by measuring the 'phase shift' of minima. The output transmission spectrum of *K* sensor units multiplexed in series is given by (Fu et al., 2009),

$$\frac{P_{output}}{P_{input}} = 10Log_{10}\prod_{k=1}^{K}\left(\frac{1}{2}L_k\left[1-\cos(\frac{2\pi}{S_k}\lambda+\theta_k)\right]\right)[\text{dB}], \tag{24}$$

where *L k, S k, ϑ k* are the loss, the period of the output spectrum and the initial phase of the *k*-th sensor, respectively. Note that the output spectrum is indeed the multiplication of individual sensor signals. PM-PCF1 (length of 20 cm) was placed freely on a table, while PM-PCF2 (length of 60 cm) was placed inside a sealed pressure chamber. Pressure was applied to PM-PCF2 from 0 – 3 bars in steps of 0.5 bar, and was measured by a pressure gauge (COMARK C9557). Fig. 19 shows the output spectra of various pressure values measured by the OSA. In principle, to obtain the sensing information, the wavelength shift of the transmission minima of each sensor needs to be determined. However, as can be seen, the multiplexed sensor signal is more complex, and so simple tracing of the initial phase may not yield accurate results. We applied the DWT to the sensor signal, and the detail coefficients at two different levels are shown in Fig. 20. These two sets of coefficients are indeed the original sensor signal of the two sensors, and from the figure, the spectrum of PM-PCF2 shifted linearly with increasing pressure, while PM-PCF1 remained unchanged (at least the initial phase). Figs. 21(a)and 21(b) show the spectral shift of the two sensors as a function of applied pressure, and the crosstalk (includes other sources of errors, such as measurement error and ambient noise) between them, respectively. Thus, this example clearly demonstrated the capability of DWT in demultiplexing and demodulating multiplexed PCF sensor signals.

6. APPLICATION CASE 4 – MEASURANDS ANALYSIS OF NOVEL FIBRE SENSORS

In the fourth example, DWT is used as a signal analysis tool to demodulate, analyse and interpret acquired signals from two novel fibre sensors, namely tilted moiré FBG and tilted Bragg reflector fibre laser (TBR-FL). These two sensors were proposed to perform simultaneous two-parameter sensing using only single sensing elements. The advantages are: (i) the capability of detecting more measurands using fewer sensors; (ii) the ability of mitigating the issue of cross-sensitivity (mostly temperature-induced) inherited from the sensor property (Rao, 1997); and (iii) sensor structure can be made more compact, which simplifies and eases the packaging and installation works for practical applications. It is known that, in order to measure two parameters simultaneously, either two distinctive sensor types or sensors of two different spectral characteristics are needed. The proposed sensors are unique in the sense that they are designed in such a way that, within a single sensor structure, their spectral characteristics response differently to different measurands, and so permitting them to distinguish individual measurand-induced changes. As a result, their spectral profiles are relatively more complex, and can be considered as having two parts from two different sensor types. Therefore, the spectra cannot be easily separated and analysed. However, with DWT, such complex sensor signals are readily separated without losing measurands information.

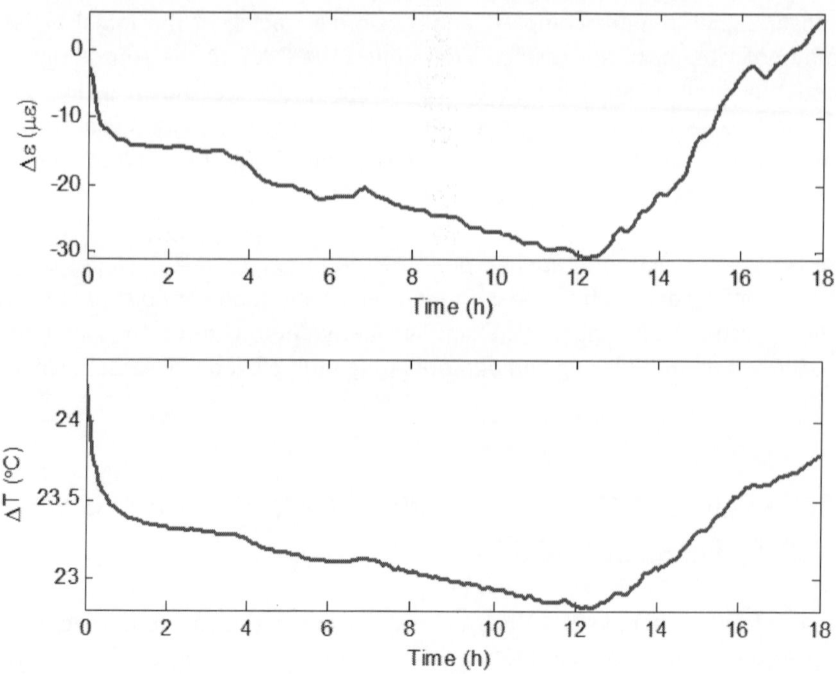

Figure 17. Simultaneous measurement of strain and temperature of an Al-alloy.

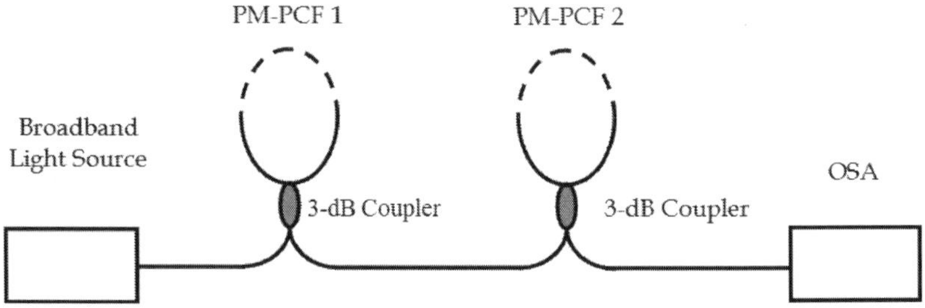

Figure 18. Experimental setup of in series multiplexing technique for PM-PCF based Sagnac interferometric sensor.

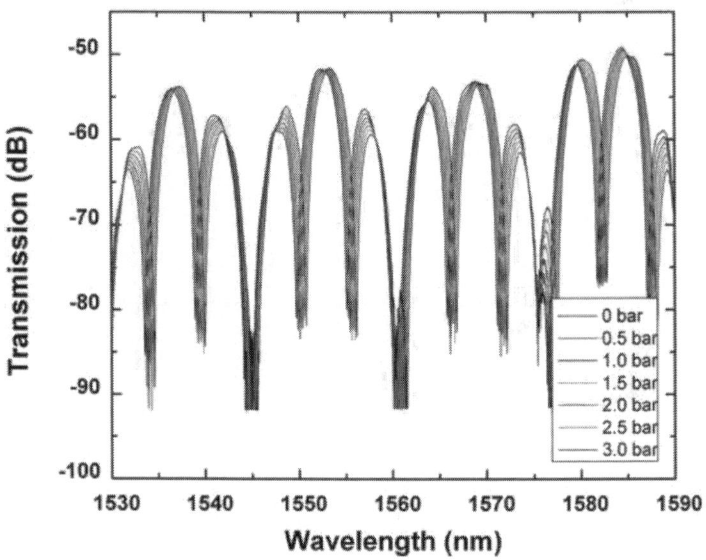

Figure 19. Output transmission spectra of the two multiplexed Sagnac interferometric sensors in series with one sensor under applied pressure variations.

6.1. Tilted Moiré Fibre Bragg Gratings

The tilted moiré FBG was originally proposed as a bandwidth controllable filter for telecommunications applications (Wong et al., 2010b). Here, we extended its application to fibre sensing, in particular, it was proposed as a single sensor for simultaneous two-parameter sensing (Wong et al., 2010c). The design criteria and fabrication procedure are detailed in (Wong et al., 2010c), and a typical spectrum is shown in Fig. 22. It can be seen that the sensor signal consists of two separate parts associated with the phase-shifted main Bragg mode (with a narrow resonance dip) and discrete cladding modes (including the ghost mode). The former is mostly confined in the core of the fibre, while the latter exist in the cladding region of the fibre. In that region, light is only loosely confined and its intensity decays exponentially along the radial direction and eventually radiates out of the fibre. As such, the cladding modes are capable of interacting with the surrounding environment at the fibre boundary, and sensing can be carried out via the principle of evanscent wave sensing.

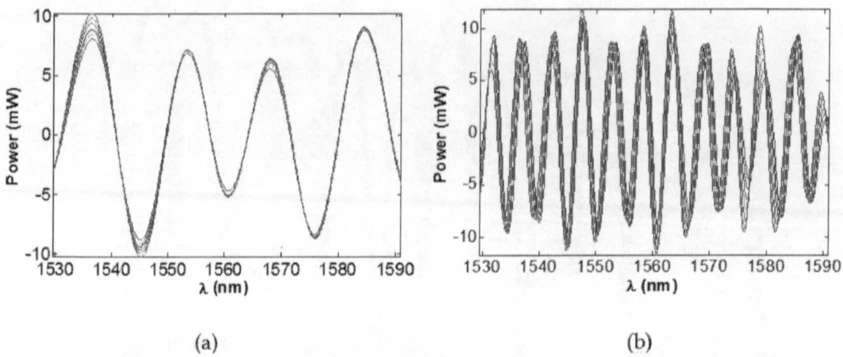

Figure 20. Detail coefficients of two PM-PCF sensors at two wavelet levels.

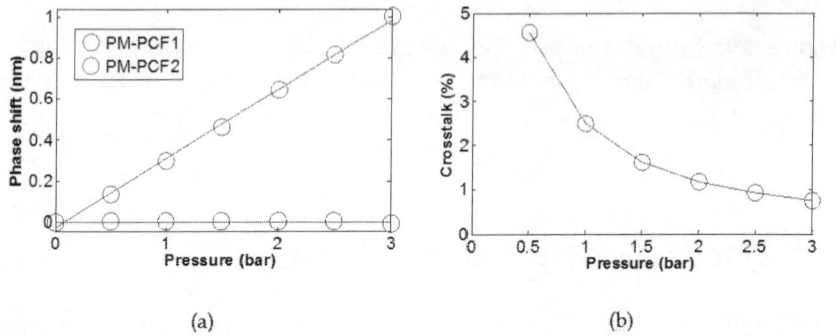

Figure 21. a) Spectral phase shift and (b) crosstalk of the two PM-PCF sensor as a function of applied pressure.

To demonstrate the use of a single sensor for simultaneous two-parameter sensing, we performed simultaneous measurement of temperature and refractive index (RI) of an aqueous solution. The aim was to obtain the measurand-induced sensitivity coefficients independently, such that the wavelength shifts from each measurand can be unambigously determined. Thus, we carried out the characterisation experiments separately to obtain the temperature and RI sensitivity coefficients. First, temperature was varied by putting the sensor into a container filled with hot water, and the value was measured by a digital thermometer (Fluke 52II with K-type thermocouple) in the range of 43°C – 77°C, captured at every 1°C intervals. Next, the sensor was placed in a container filled with pre-mixed and saturated glucose solution (Dextrosol D-Glucose powder), and RI was varied by adding water to dilute the solution. A digital refractometer (Reichert AR200) was used to obtain the absolute RI and temperature of the samples. The measured RI range was between 1.3331 and 1.4117, set by the RI of water and that of saturated glucose solution. The temperature (obtained from the refractormer) varied between 21.0°C – 21.6°C during the experiment. In both cases, wavelengths of the Bragg and cladding modes shifted according to measurands changes. More specifically, temperature changed the Bragg wavelength via thermo-optic effect (thermal expansion of glass fibre is very small that can be neglected), whereas the RI changed both the wavelength and transmission loss of the cladding modes. Thus, by applying the DWT, these two types of modes can be separated and their changes due to the two measurands can be extracted and analysed. Fig. 23 shows the detail wavelet coefficients for the (a) Bragg mode and (b) cladding modes at the 7th- and 6th-levels, respectively. Since the detail coefficients are mean removed, and so in the Bragg mode, the narrow resonance dip becomes a sharp peak, and temperature change can be measured by tracking this peak shift. As for the cladding modes, it is known that each individual cladding mode responds differently to the same measurand (e.g., RI) change, and taking an average of a number of cladding modes shifts will give more accurate readings than tracking just one particular mode. RI is thus measured by the averaged wavelength shift of about 10 cladding modes.

After taking the DWT to the measured spectra, Fig. 24 (a) shows the wavelength shifts of the wavelet coefficients for the Bragg and averaged cladding modes, respectively, as a function of temperature. The solid lines are the linear regression fits of the data points. It is clear that, in response to temperature change, the averaged cladding modes shifted by almost the same amount and in the same direction as the Bragg mode. From the regression lines, the temperature-induced sensitivity of the Bragg and averaged cladding modes are 10.53 pm/°C (with R^2 = 0.9979) and 10.48 pm/°C (with R^2 = 0.9925), respectively, which correlated very well with each other. Thus, it is sufficient to measure the temperature-only change by tracking the wavelength shift of the Bragg mode. Similarly, Fig. 24 (b)shows the wavelength shifts of the wavelet coefficients for the Bragg and averaged cladding modes, respectively, as a function of RI. From the figure, the Bragg mode remained unperturbed (within measurement errors), indicated that it is insensitive to RI change. The cladding modes, however, varied nonlinearly, with the curve best fitted by a polynomial function given empirically by,

$$P_{DWT}(n) = 169.58n^4 - 655.26n^3 + 843.59n^2 - 361.85n \tag{25}$$

(with R^2 = 0.9987) for the specified range $1.3331 \leq n \leq 1.4117$. Thus, for RI-only change, Eq. (25)can serve as a lookup table, such that by tracking the wavelength shift of the averaged cladding modes, the RI value can be obtained. By combining these two distinct measurands-induced responses, simultaneous measurement of temperature and RI is realised. Temperature change can be directly obtained from the wavelength shift of the Bragg mode, and the RI from the differential wavelength shifts between the Bragg and averaged cladding modes, i.e., the residual amount after subtracting the average shifted amount of the cladding modes from that of the Bragg mode. Fig. 24(b) (and Eq. (25)) can then be used to find the RI value by looking-up the corresponding measured differential shift.

6.2. Tilted Bragg Reflector Fibre Lasers

Up to present, there are mainly two types of grating-based FLs, namely distributed feedback and distributed Bragg reflector FLs. When applied in sensing, FLs have the advantage of high sensitivity, sensing output power and extinction ratio, and narrow linewidth/bandwidth. For simultaneous two-parameter sensing using single sensors, thus far, most of the proposed works are of passive type and only very few on using active sensors, e.g., fiber Raman lasers (Han et al., 2005, Tran et al., 2005) and distributed feedback FLs (Hadeler et al., 1999, 2001). TBR-FL is a new type of FL formed by a pair of wavelength and tilt-angle matched tilted FBGs (TFBGs) (Wong et al., 2011). In addition to the lasing peak, it possessed a grating tilt-induced cladding modes spectrum, which provided an extra sensing mechanism to detect the surrounding environment. We demonstrate that, with a simple experimental setup, the use of a single TBR-FL for simultaneous sensing of temperature and RI.

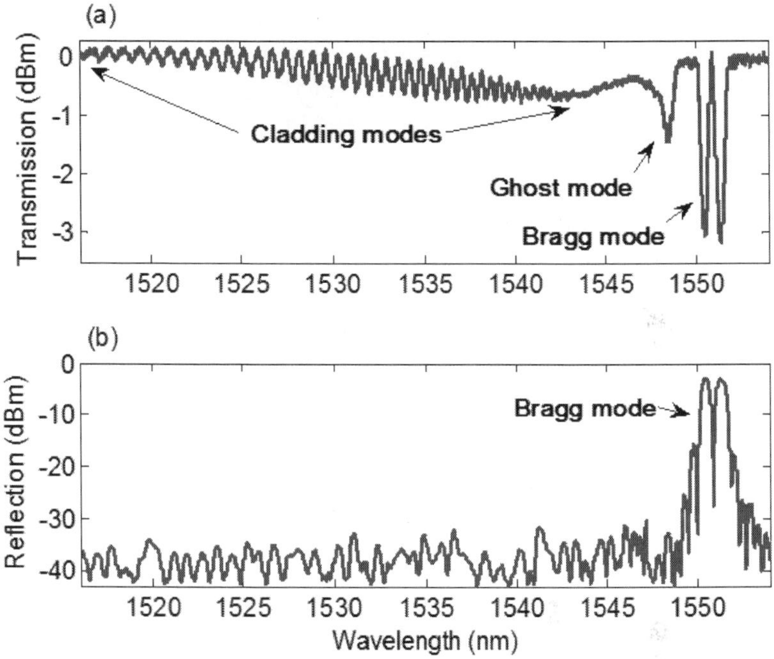

Figure 22. a) Transmission and (b) reflection spectra of the TMFBG sensor for simultaneous two-parameter sensing.

The structure of the TBR-FL comprises a pair of wavelength and tilt-angle matched TFBGs forming the resonant cavity and is depicted in Fig. 25(a). The fabrication procedure is described in (Wong et al., 2011). The experimental setup depicted schematically in Fig. 25 (b). A laser diode was pumped to the TBR-FL via a wavelength-division-multiplexer (WDM), and the output signal was obtained by the OSA. At the other side of the FL was a continuous piece of EDF (~50 cm) looped in a diameter of ~3 cm, and so the sensor head comprised both the FL and the short coiled EDF section. Index matching gel (IMG) was applied to the far end of the EDF to minimise any reflections that may cause resonant feedback. This EDF section acted as an ASE source when excited by the excessive pump source. As such, the transmission spectrum of the constituent TFBG pair can be observed. With this setup, both the laser output and cladding modes spectra can be obtained simultaneously. A typical full spectrum is shown in Fig. 26(a), which consisted of both the laser output [Fig. 26(b)] and cladding modes [Fig. 26(c)] spectra. Lasing occurred at the Bragg mode bound inside the core, whereas cladding modes were coupled out from the core and did not contribute to the laser operation. As mentioned, cladding modes can interact with the surrounding environment via evanescent wave, and therefore used to perform RI sensing. On the other hand, temperature change affected the FL as a whole and altered the entire spectrum (including lasing and cladding modes). Thus, simultaneous sensing was achieved by combining these two sensing properties, i.e., temperature was measured by tracking the lasing mode shift, and RI by the differential wavelength shift between the laser output and cladding modes. Similar to the previous case, the main objective was to empirically obtain the temperature and RI induced sensitivity coefficients. As such, characterisation experiments similar to that for tilted moiré FBGs were conducted [see also (Wong et al., 2011) for details], and the DWT was employed to demodulate and analyse the acquired sensor signals. Fig. 27 shows the wavelet coefficients of a typical measured spectrum. Detail coefficients are extracted for cladding modes [Fig. 27(a)], as for the previous case; whereas approximation coefficients are extracted for the lasing mode [Fig. 27(b)]. Since we are only interested in the lasing wavelength shift, approximation coefficients represent and resemble the original signal more accurately. After taking the DWT to the measured spectra, wavelength (i.e., wavelet coefficients) shifts of the laser output and averaged cladding modes (~20 modes) as a function of temperature are shown in Fig. 28(a). Solid lines are the linear regression fits of the data. From the figure, both the laser output and averaged cladding modes yielded a very similar temperature-induced sensitivity, having values of 10.75 pm/°C (R^2 = 0.9991) and 10.88 pm/°C (R^2 = 0.9975), respectively. With such a high degree of correlation, it is sufficient to measure the temperature change by tracking the lasing wavelength shift alone. For RI, Fig. 28(b) shows the wavelength shifts of the laser output and averaged cladding modes as a function of RI value. The laser output remained at the zero value (within measurement errors) and so have no direct relationship with RI. The averaged cladding modes varied nonlinearly with RI, and within the measured range the empirical relationship can be described by a polynomial function,

$$\Delta\lambda_{Cladding}(n) = 358.2n^4 - 1432.1n^3 + 1910.1n^2 - 849.9n \quad (26)$$

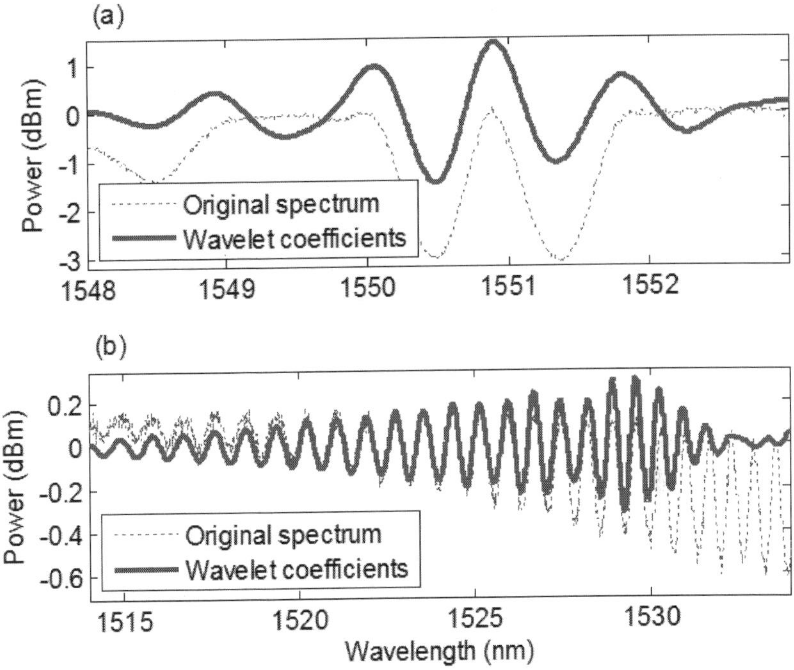

Figure 23. Wavelet coefficients (solid lines) of a measured transmission spectrum: (a) 7th-level coefficients for the Bragg mode, and (b) 6th-level coefficients for cladding modes. Dotted lines are the original spectra.

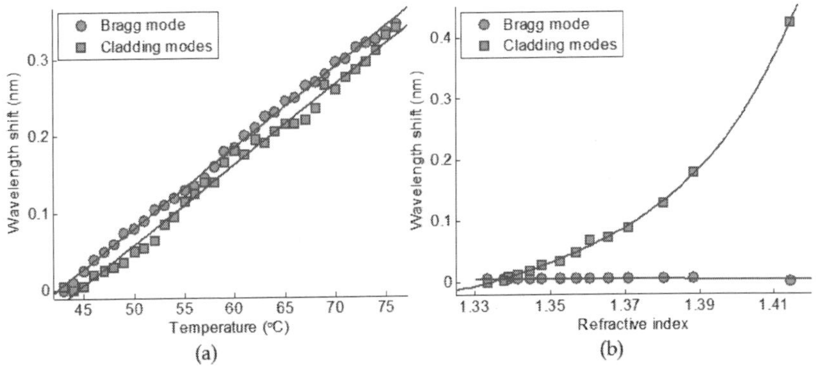

Figure 24. Wavelength shifts of the wavelet coefficients for the Bragg mode and averaged cladding modes as a function of (a) temperature and (b) refractive index. Lines are the regression fits.

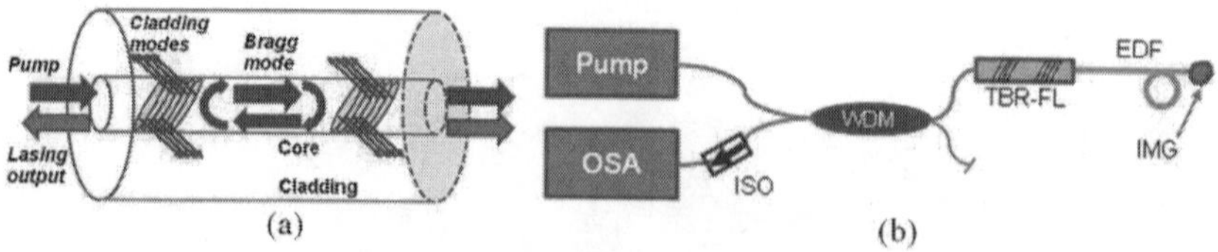

Figure 25. a) Structure of the TBR-FL, and (b) experimental setup. OSA = optical spectrum analyser, WDM = wavelength division multiplexer, EDF = erbium-doped fibre, ISO = isolator, IMG = index matching gel.\

($R^2 = 0.9998$). Thus, RI-only change can be obtained from the wavelength shift of the averaged cladding modes, as the laser output was not sensitive to it. Having established the temperature and RI sensitivity coefficients, simultaneous sensing of these two measurands using a single TBR-FL can be achieved. Temperature change can be obtained from the wavelength shift of the laser output, whereas RI change from the differential wavelength shift between the averaged cladding modes and laser output. That is, the residual amount of the wavelength shift of the averaged cladding modes after subtracting from that of the laser output. As such, Fig. 28(b) can be used as a look-up table to find the RI from the differential wavelength shift.

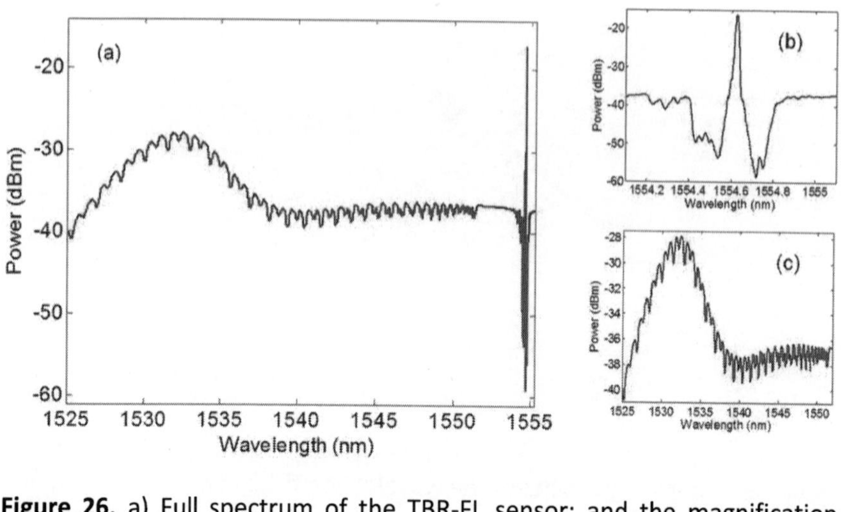

Figure 26. a) Full spectrum of the TBR-FL sensor; and the magnification around (b) the laser output and (c) cladding modes spectrum.

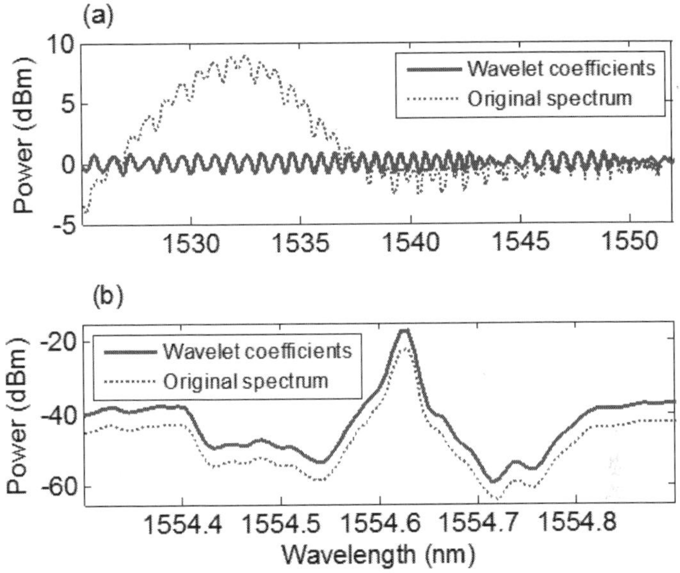

Figure 27. Wavelet coefficients of a measured TBR-FL spectrum: (a) 6th-level detail coefficients for the cladding modes, and (b) 2nd-level approximation coefficients for the Bragg mode. Dotted lines are the original spectra (manually offset) for comparison.

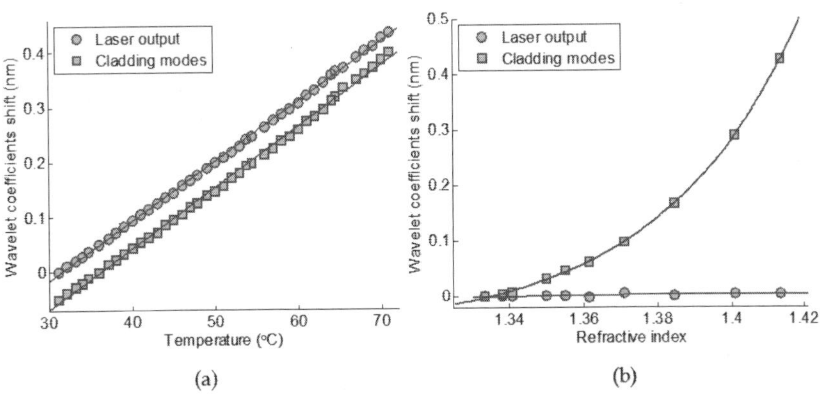

Figure 28. Wavelet coefficients shift of the laser output and averaged cladding modes as a function of (a) temperature and (b) refractive index. Lines are the regression fits.

7. CONCLUSION

A review of the applications of DWT in optical fibre sensing is presented. Several representative application examples proposed by the authors have been discussed; and based on the implementation of DWT, novel signal processing techniques and fibre sensors have been designed and proposed. The concepts of DWT demodulation technique and the BLT wavelet denoising were introduced and applied to various application examples. First, we proposed and employed the DWT to demodulate and demultiplex a multiplexed FFI and FBG sensor system. Second, we designed a novel type of FBG (the AMCFBGs), and based on their unique overlapping properties, a new multiplexing technique called spectral overlap multiplexing was proposed and demonstrated. Third, DWT was employed in the multiplexing and demultiplexing of PCF-based sensors array, a relatively new class of fibre sensors that, up until now, has always been used as single sensors. Fourth, DWT was used as a signal analysis tool for two novel fibre sensors, namely the tilted moiré FBG (a passive sensor) and the TBR-FL (an active sensor) sensors. Complex sensor signals were separated and demodulated to obtain the individual measurands-induced changes, such that single sensing elements can perform simultaneous two-parameter sensing. In addition, BLT was applied in all the above cases to denoise sensor signals automatically during the use of the DWT demodulation technique.

ACKNOWLEDGEMENTS

Funding for an ARC Linkage Project (Project ID: LP0884100) from the Australian Research Council and the Roads and Traffic Authority, NSW, Australia, is gratefully acknowledged. Funding for an International Science Linkage Grant (Project ID: CG130013) from the Department of Industry, Innovation, Science and Research (DIISR), Australia, is also gratefully acknowledged.

REFERENCES

1. H. J. Bang, C. G. Kim, 2010Study on the wavelet decomposed details of impact induced AE signals in composite laminates using fiber Bragg grating sensors, Proc. SPIE, 7647 76473E

2. C. C. Chan, N. Ni, J. Sun, Y. C. Chu, Y. Tang, C. L. Poh, 2007Improving the detection accuracy in fiber Bragg grating sensors by using a wavelet filter, J. Optoelectron. Adv. Mater., 9 8 2376 2379

3. C. C. Chan, N. Ni, J. Sun, Y. C. Chu, Y. Tang, C. L. Poh, 2010Interferometric noise suppression in fiber Bragg grating sensors by using wavelet filter, J. Optoelectron. Adv. Mater., 12 6 1241 1246

4. P. Childs, A. C. L. Wong, B. B. Yan, M. Li, G. D. Peng, 2010A review of spectrally coded multiplexing techniques for fibre grating sensor systems, Meas. Sci. Technol., 21 094007pp)

5. I. Daubechies, 1992 Ten Lectures on Wavelets, Society for Industrial and Applied Mathematics, Philadelphia, PA, U. S. A.

6. D. L. Donoho, I. M. Johnstone, 1994Ideal spatial adaption by wavelet shrinkage Biometrika, 81 425 455

7. H. Y. Fu, A. C. L. Wong, P. A. Childs, H. Y. Tam, Y. B. Liao, C. Lu, P. K. A. Wai, 2009Multiplexing of polarization-maintaining photonic crystal fiber based Sagnac interferometric sensors, Opt. Express, 17 18501 18512

8. T. K. Gangopadhyay, S. Chakravorti, K. Bhattacharya, S. Chatterjee, 2005Wavelet analysis of optical signal extracted from a non-contact fibre-optic vibration sensor using an extrinsic Fabry-Perot interferometer, Meas. Sci. Technol., 16 1075 1082

9. T. K. Gangopadhyay, S. Chakravorti, S. Chatterjee, K. Bhattacharya, 2006Time-frequency analysis of multiple fringe and nonsinusoidal signals obtained from a fiber-optic vibration sensor using an extrinsic Fabry-Perot interferometer, J. Lightw. Technol., 24 5 2122 2131

10. O. Hadeler, E. Rønnekleiv, M. Ibsen, R. I. Laming, 1999Polarimetric distributed feedback fiber laser sensor for simultaneous strain and temperature measurements, Appl. Opt., 38 1953 1958

11. O. Hadeler, M. Ibsen, M. N. Zervas, 2001Distributed-feedback fiber laser sensor for simultaneous strain and temperature measurements operating in the radio-frequency domain, Appl. Opt., 40 3169 3175

12. Y. G. Han, T. V. A. Tran, S. H. Kim, S. B. Lee, 2005Multiwavelength Raman-fiber-laser-based long distance remote sensor for simultaneous measurement of strain and temperature, Opt. Lett., 30 1282 1284

13. K. J. Jones, 2000aWavelet edge detection applied to be used with composite embedded fiber optic sensors, Proc. SPIE, 3986 397 406

14. K. J. Jones, 2000bWavelet crack detection algorithm for smart structures, Proc. SPIE, 4328 306 313

15. H. Lamela-Rivera, C. Macia-Sanahuja, J. A. Garcia-Souto, 2003Detection and wavelet analysis of partial discharges using an optical fibre interferometric sensor for high-power transformers, J. Opt. A-Pure Appl. Opt., 5 1 66 72

16. B. Lee, 2003Review of present status of optical fiber sensors, Opt. Fiber Technol., 9 57 79

17. S. G. Mallet, 1998 A Wavelet Tour of Signal Processing, Academic Press, Burlington, MA, U. S. A.

18. S. G. Mallet, 1989A theory for multiresolution signal decomposition: the wavelet representation, IEEE Trans. Pattern Anal. Mach. Intell., 11 674 693

19. Y. J. Rao, 1997In-fibre Bragg grating sensors, Meas. Sci. Technol., 8 355 375

20. Burrus. C. Sidney, R. A. Gopinath, H. Guo, 1998 An Introduction to Wavelets and Wavelet Transforms, Prentice Hall, Upper Saddle River, NJ, U. S. A.

21. W. J. Staszewski, Pierce. S. Gareth, K. Worden, W. R. Philp, G. R. Tomlinson, B. Culshaw, 1997Wavelet signal processing for enhanced Lamb wave defect detection in composite plates using optical fiber detection, Opt. Eng., 36 7 1877 1888

22. M. Tomic, Z. Dinovic, S. Petricevic, 2010Fiber-optic pressure sensor based on Fizeau receiving interferometer, Proceedings of the 33rd International Convention MIPRO, 100 104Opatija, Croatia, May 24-28, 2010

23. T. V. A. Tran, Y. G. Han, S. H. Kim, S. B. Lee, 2005Long-distance simultaneous measurement of strain and temperature based on a fiber Raman laser with a single fiber Bragg grating embedded on a quartz plate Opt. Lett., 30 1632 1634

24. M. Vetterli, J. Kovacevic, 1995 Wavelets and Subband Coding, Prentice Hall, Upper Saddle River, NJ, U. S. A.

25. G. Wang, K. Pran, G. Sagvolden, G. B. Havsgard, A. E. Jensen, A. E. , G. A. Johnson, S. T. Vohra, 2001Ship hull structure monitoring using fibre optic sensors, Smart Mater. Struct., 10 472 478

26. A. C. L. Wong, C. J. E. Phillips, G. D. Peng, 2005Improved demodulation algorithm for spatial-frequency multiplexed fibre-optic Fizeau strain sensor system, Proceedings of 30th Australian Conference on Optical Fibre Technology, Sydney, Australia

27. A. C. L. Wong, P. A. Childs, G. D. Peng, 2006aSimultaneous demodulation technique for a multiplexed fibre Fizeau interferometer and fibre Bragg grating sensor system, Opt. Lett., 31 23 25

28. A. C. L. Wong, P. A. Childs, G. D. Peng, 2006bMultiplexed fibre Fizeau interferometer and fibre Bragg grating sensor system for simultaneous measurement of quasi-static strain and temperature using discrete wavelet transform, Meas. Sci. Technol., 17 384 392

29. A. C. L. Wong, P. A. Childs, G. D. Peng, 2007aMultiplexing technique using amplitude-modulated chirped fibre Bragg gratings, Opt. Lett., 32 1887 1889

30. A. C. L. Wong, P. A. Childs, G. D. Peng, 2007bSpectrally-overlapped chirped fibre Bragg grating sensor system for simultaneous two-parameter sensing, Meas. Sci. Technol., 18 3825 3832

31. A. C. L. Wong, P. A. Childs, G. D. Peng, 2010aSpectrally coded multiplexing techniques in fibre-optic sensor systems, In: Trends in Photonics, J. Canning (ed.), Chapter 8, 233 276Transworld Research Network, 978-8-17895-441-7Kerala, India

32. A. C. L. Wong, M. Giovinazzo, H. Y. Tam, C. Lu, G. D. Peng, 2010bTilted moiré fiber Bragg grating optical filters with controllable passband and stopband, J. Lightw. Technol., 18 898 904

33. A. C. L. Wong, M. Giovinazzo, H. Y. Tam, C. Lu, G. D. Peng, 2010cSimultaneous two-parameter sensing using a single tilted moiré fiber Bragg grating with discrete wavelet transform technique, IEEE Photon. Technol. Lett., 22 1574 1576

34. A. C. L. Wong, W. H. Chung, C. Lu, H. Y. Tam, 2011Single tilted Bragg reflector fiber laser for simultaneous sensing of refractive index and temperature, Opt. Express, 19 409 414

Index

S

T

W